Bryan Gets it!

This is a must read for any law enforcement officer wanting to be SWAT team member, those currently serving on a SWAT team and those in command, to include every Police Chief and Sheriff that has a SWAT team in their department.

De Re SWAT covers it all, from the all important fundamentals, to explaining the complexities of fast paced SWAT operations in a highly charged environment. It is written by an expert operator who's been on the front lines as both a SWAT officer and then team supervisor. Bryan knows getting it right every time is a product of dedicated training, discipline and knowledge, not luck.

Chiefs and Sheriffs, applying the principles in *De Re SWAT* will make your teams better, keep your officers and citizens safer, and help keep you out of court and the harsh glare of an unfriendly media spotlight.

As the Chief of Police of the fifth largest city in California and as a former SWAT team member and SWAT sergeant, I salute Bryan for his incredible knowledge and experience, and his willingness to share it with the law enforcement community to help take SWAT to the next level of professionalism. Well done!

—Jerry Dyer, Chief of Police
Fresno, California
Former President, California
Chiefs of Police Association.

I'm overwhelmed by the depth and detail of this book. It is a masterpiece, patterned after a true, timeless classic, with tremendous value for operators and leaders.

—David Grossman
Author of *On Killing* and *On Combat*
www.warriorsciencegroup.com
www.killology.com

"...this comprehensive book provides an overview of contemporary SWAT operations and will prove valuable to command staff and supervisory personnel. The manner in which Bryan Morgan models the book after the 19th Century military classic "De Re Militaria" is very creative and useful for the modern-day tactical arena..."

—Captain Rick Book
Sacramento Sheriff's Department
Former SED Team Leader
California Peace Officers' Association

DE RE SWAT

To Charlene Cornell

It was my honor to train with you. I have no doubt that you will have a successful career. I'd work with you anytime. I can see your quality and professionalism. Keep the faith.

Byron [illegible]

DE RE SWAT

BRYAN MORGAN

To Brandon

I hope this may serve you in your career. Be safe and keep the faith

Bryan M

TATE PUBLISHING & *Enterprises*

Published by Tate Publishing & Enterprises, LLC
127 E. Trade Center Terrace | Mustang, Oklahoma 73064 USA
1.888.361.9473 | www.tatepublishing.com

Tate Publishing is committed to excellence in the publishing industry. The company reflects the philosophy established by the founders, based on Psalm 68:11,
"The Lord gave the word and great was the company of those who published it."

Edited by Vince Conn
Cover design by Amber Lee
Interior design by Jeff Fisher

Published in the United States of America

ISBN: 978-1-60799-651-4
1. Reference, Handbooks & Manuals
09.08.10

For Tracy,
my beautiful wife
and
the love of my life

TABLE OF CONTENTS

Chapter 2. Organization of the SWAT Establishment

Part 2. De Re Militari

Chapter 4. The Selection and Training of New Levies

Chapter 5. The Organization of the Legion

Chapter 6. Dispositions for Action

INTRODUCTION

To the reader,

This book was authored to reveal the proper selection, operations, and maintenance of SWAT teams. It is aimed at police command staff and supervisory personnel, but would also be of great utility to newer officers. The field of SWAT is of extreme interest to the general public and the military. The book addresses all facets of the SWAT establishment, including purpose, deployment, special duties, incident command systems, crisis negotiations, and actual decision-making criteria as related to resolving emergency crises. I modeled and organized the book after the famous military manual "De Re Militaria" written by the Roman General Vegetius in the fourth century. Vegetius' purpose was to educate his emperor on the role and function of the Roman Army and to describe its general inner workings.

I am an honorably retired deputy sergeant (due to an on-the-job injury). I worked for the Fresno County Sheriff's Department in Fresno, California, for eighteen years as a sworn officer. I was assigned as a SWAT officer for ten years and as a SWAT supervisor for an additional three years. After my injury, I served as an intelligence specialist with the US Department of Justice assigned to the Central Valley High Intensity Drug Trafficking Agency.

During my time in SWAT, I served on over one hundred and twenty SWAT actions. I was also selected to serve on two California Police Officers' Standardized Training Commissions. Those commissions explored and recommended standardized training courses for SWAT personnel. I have received commendations at the local, state, and federal levels for my performance in law enforcement.

I have served as an instructor with the Fresno City College Police Academy and for the Fresno County Sheriff's Department. I was issued basic, intermediate, advanced, and supervisor's policing certificates by the California POST commission. I also obtained instructor or trainer certificates as a range master, interview and interrogation instructor, rappel master, sub-machine gun instructor, carbine instructor, field training officer, and incident command systems and emergency management trainer.

I served my community, my state, and my country. I enforced and honored its laws and constitution. I have personally stepped in harm's way and ordered others into harm's way to serve and protect. It is my hope that if this book can serve as a reference or introduction for SWAT, I still might serve my community.

Very truly yours,
Bryan Morgan
"30 SECONDS!"

PREFACE

In the fourth century a Roman military officer named Vegetius feared for the safety of Rome because of the decline of her military establishment. Vegetius felt so impassioned about the decline of the Roman military that he authored three letters to the Emperor Valentinian. The goal of Vegetius' writings was to educate his emperor not only of the role and function of the Roman Army, but also to describe in general its inner workings, importance to the survival of Rome, and the reasons for Rome's military successes and alternately its current decline. Vegetius' letters were extremely appropriate and found to be so valuable they remain alone as surviving military manuals of that period. Vegetius took quite a risk writing these open letters to his emperor. The penalty for insult would have been death. Vegetius believed so much in the need for his profession he risked all to impart his knowledge.

As a young SWAT operator years ago, I recognized there were some command staff in my department that had very little knowledge of the SWAT establishment's function, purpose, and utility. I found there was no one brave enough to step forward as Vegetius had done. I later became a supervisor and saw this lack of knowledge about the SWAT establishment wasn't confined to my department; in fact, practically every other agency had some command staff with little or no practical knowledge of the art of SWAT. So with a purpose very much the same as Vegetius', I submit his works translated to today's world and the art of SWAT.

Today's policing agencies must respond to a complicated and ever-changing world of escalating violence. The prying eyes and ears of the media report every incident and then pass summary judgment. Failure to respond appropriately to these incidents frequently leads to unnecessary injuries, deaths, civil liability, the downfall of the administration responsible, and ultimately, the loss of the public's trust. While I do not share the risk of loss of life that Vegetius did in his writings, I do place myself in position for ridicule by my peers.

The importance of this translation is thus to be remembered; the violent criminals and terrorists that police departments will encounter need not be good at what they do, they need be lucky just once to succeed. The officers and SWAT operators do not have that luxury. They must be very, very good each and every time they respond to an incident, and they can never rely upon luck. If the right people are not in the right positions, or if those people haven't been trained and disciplined properly, the odds of failure increase unnecessarily. This failure would not be their fault. The true responsibility would lie with the superintendents of the agency involved.

PART 1:

DE RE SWAT

1

SELECTION AND TRAINING OF THE SWAT ESTABLISHMENT

Preface to Chapter 1—To My Sheriff/Chief

It has been a custom for anyone involved in police training to offer command staff the fruit of their studies in written form. In fact, no work can be published by an officer with propriety but under the auspices of their superintendents. The command staff's knowledge must be more general, and of the most important kind, as their influence is felt so keenly with the rest of his department. There are many instances of favorable receptions from command staff conferred on the works presented to them, and this encouragement makes the science of policing in general flourish. The consideration of indulgence for attempts of this source induced me to follow this example.

I have not forgotten my own inability and inexperience when compared with experts from busier and more experienced departments. One advantage, however, I derive in being able to write about this field is it requires no elegance of expression or extraordinary share of genius. I need only take great care in the fidelity of the collection and explanation of these works. These works must also meet the same high standard of public scrutiny every policing agency is held to in its training and actions. Finally this work must also be able to stand up to the media's reporting of it. Every department's performance is highly affected by the media, and many of the changes made in policing today are precipitated by their reporting of police performance.

My design in this treatise is to exhibit in some order the peculiar customs and usages of the better departments in the choice of new SWAT operators. I do not presume to offer this work from a supposition the reader is not acquainted with every part of policing, it will be seen that the same policy and procedures that were established for the welfare of the department and its community are formally observed by the SWAT establishment, and this abridgement may also be used as a reference on a very necessary and important subject.

The SWAT Team's Discipline Is the Cause of Their Greatness

A successful SWAT action does not depend upon mere numbers of operators or even upon the operators' individual willingness to engage and arrest violent criminals; only skill and discipline will ensure a successful SWAT action. It has been found that departments with longstanding successful SWAT teams owe that success to no other cause than continual training, exact observance of discipline, and unwearied cultivation of the art of special weapons and tactics as related to policing. Without these, what chance would the inconsiderable numbers of SWAT teams truly have to successfully solve the prodigious numbers of call-outs they have experienced? What of their actions against barricaded suspects? How else have they always been successful in the service of search warrants against fortified crack houses? What about the sheer numbers of violent street gang members they routinely arrest? The street gang members not only frequently surpass the entire agency in numbers, but sometimes in armament as well. The SWAT establishment will always be unequal in the ability to indiscriminately respond with violence, thus they must succeed by usage of discipline, deception, and stratagem. In light of today's media response to every police action and use of force, the SWAT establishment must be indisputably superior in their skills and reasonable application of force.

The entire SWAT establishment must take unusual care in the choice of its team members and in their training to use every advantage. They must thoroughly understand the importance of hardening by continual practice and of training in every maneuver and be ready for everything that might happen in the line of their duties. The SWAT establishment must also be strict in punishing idleness and sloth. The courage and skill of the operators is heightened by their knowledge of their profession, and the SWAT establishment should only want an opportunity to execute what they are convinced they have been perfectly taught. A handful of operators, properly supported and competently led, inured to the uses of force, proceed to certain victory, while on the contrary, numerous teams composed of raw or undisciplined officers are but multitudes headed toward disaster, usually in a wrongful application of force that results in unnecessary injuries or death and a subsequent civil lawsuit.

The Selection of Recruit Operators

In order to treat this subject with some method, we will first examine which departmental officers are to be preferred for supplying the SWAT team with new operators. It is certain that every officer in an agency has the ability to be a SWAT operator, as all of the officers are brave, but it is equally as certain that some of the officers will be more prone to make good SWAT operators. They will be able to make decisions under stress, carry the burdens and heavy load of standard-issued SWAT equipment, as well as being physically strong and well conditioned. This will allow them to perform their duties at all times and in all parts of the jurisdiction.

Should younger or older officers be selected for the SWAT team? Which are the best candidates to be SWAT operators? There is no doubt that the younger officers are more likely to be the most fit to carry and use the arms and stand up to the physical rigors of SWAT training and SWAT actions. The younger officers will have had fewer injuries and generally be in better physical condition. They will be able to bear the standard SWAT equipment load in the greatest heat of the day, will most likely be unacquainted with the desk jobs of an agency, and are typically more than willing to work physically. They are also simpler, have fewer bad habits, find the assignment attractive, and are more likely to still have the discipline and physical conditioning from the police academy or prior military service. They will be more able to carry additional equipments and climb barriers, as well as carry the weight of an operator's standard-assigned equipment. In cases of necessity it may be necessary to take officers from the specialty units and older officers. In these cases, those officers should be thoroughly tested physically to make sure they can carry the loads necessary, and have the physical endurance required. Simple participation in SWAT training itself should not lead to an injury that ends their career. While SWAT operators should be fit, they need not be weightlifters, martial arts experts, or marathon runners. The goal is to instruct them in the perfect use of their firearms and intermediate force weapons. The operators must be able to not only properly use their weapons; they must be able to be deployed for extended periods, in all manner of weather, in all parts of the jurisdiction. They must have their minds trained as well as receive physical training. Only this combination will prepare them for the service they will provide.

The SWAT team and the rest of the SWAT establishment must be drawn from all parts of the department and must reflect its gender and racial makeup. However, it is unwise to pick officers and use a quota system with the intent to make sure the team is composed of the right numbers of minorities. The team's "speed"—namely its ability to perform—will be the speed and ability of its slowest or lowest-performing member. The team will be only as good as its weakest member. Select the right operators for the SWAT team and the right officers for the rest of the SWAT establishment. Do not burden the SWAT team with officers who are unable to perform the duties they will be assigned. The chief strength of the SWAT team will be recruited from the younger officers. Recruit the right people into the department to accurately reflect and serve the community, and then the department will possess the right candidates for its SWAT.

The Proper Age for Recruits

It has been the practice of successful SWAT teams for the proper time of enlisting an operator onto the team as being after their completion of probation. At this time the officers are more likely to embrace the job of a SWAT operator and policing as a whole, and they will be ready to have the training imprinted upon their minds. Besides this, the indispensable exercises of running, self-defense, range training, and use of intermediate force options (all performed while carrying a SWAT officer's full load of equipment) will be acquired before they have become hindered by advancement of their age. It is activity and continual practice that forms a useful and good SWAT operator. The best operators are inducted early in their careers and trained in the strictest manner in all the fatigues and exercises of SWAT duties. It is certainly better that an officer be perfectly disciplined but still of a proper age for SWAT action, than for an officer to have the mortification of knowing too late, at a critical moment, that their time is past. A sufficient time is also required for individual operator instruction in the different branches of SWAT service. It is no easy matter to train a sniper or scout; to teach an entry officer in every part of the dynamic entry and slow, methodical search techniques; to teach them all how to communicate; to move with stealth; to properly aim, load, and reload long arms with precision; to use less lethal weapons, use gas launchers, deploy distraction devices, operate in gas environments; to use physical force; to breach doors and windows; to make decisions under stress; and to make arrests with a proper and minimum use of lawful force. A SWAT operator who is properly trained will not shrug from an opportunity to show their ability. In fact they will be eager for an opportunity to show their abilities.

Their Size

It has been found that all ages of officers in a department can perform parts of SWAT duties, but what must be remembered is that there are certain tasks the SWAT operators absolutely must be able to perform. The operators must be able to scale six-foot walls in full SWAT gear, they must be able to run an obstacle course of one half mile in less than four minutes, and they must be able to perform at least three pull-ups in full gear. These requirements must be kept up at all times and for all operators. It is not height or ability to lift weights, the ability to run long distances, or even athletic dexterity that is highly regarded in a SWAT officer. Deficiency in any of these fields can be compensated for by dedicated operators with vigor and courage. The operators need not be big or small; they need to be able to perform the essential duties of a SWAT operator.

Signs of Desirable Qualities

Those employed to superintend new SWAT operators and officers in the SWAT establishment should be particularly careful in examining the backgrounds of candidates as well as their physical attributes. This will enable them to form a true judgment and choose those most likely to prove good operators. Experience has proven that there are certain signs by which these virtues may be discovered. The best candidates ought to have had excellent evaluations and should be known to other police officers as being trustworthy and competent officers of unquestioned integrity. The officers who can pass all of the physical tests as well as a thorough in-house background will have the best potential for developing into the best operators.

The Rest of the SWAT Establishment

When choosing the best officers for the SWAT team, you should not fail to make sure that the right people are in the positions of the command post and on the crisis negotiations teams. These officers need not be as physically fit, but should be of no less moral character than the operators. It is the very essence of a SWAT action that these parts of the SWAT establishment are interconnected and must be entrusted indiscriminately; you must have absolute fidelity in these positions too. SWAT operators are only as good as the logistical tail and support services that enable them to function and solve the event. You do not want your SWAT operators using force if possible, thus the selection of crisis negotiators is equally as important. The successful resolution of every SWAT call depends upon the inner workings of the entire SWAT establishment. All members of the SWAT establishment should have a sense of honor that prevents them from acting undisciplined; this will make them successful.

Nothing good can be expected from an operator who is a coward or sluggard even though the operator may have served on many missions and completed many training sessions. A team formed without proper regard for the quality of its members will ultimately fail. It has been proven through fatal experience that poor training and improper selection of SWAT operators leads to misfortune. Many of those failures can be imputed to the effects of long periods of inaction that tend to make the department negligent and careless of the choice of recruit operators, and to the inclination prevalent among some command staff who prefer to spend limited budgets on things other than SWAT. It is also shameful conduct for a superintendent who, through interest or connivance, allows SWAT missions to be accepted and performed by detective or specialty units that those units are not trained or equipped for. In effect, this practice sends sworn personnel on calls they cannot safely resolve. Thus this trust is of such importance that it should be committed to none but those of merit and integrity.

Certification

The new operators, crisis negotiators, and command post personnel should all receive basic certified skills training in their discipline immediately after they are recruited. This will ensure they are fit for further training. For the operators it will ensure they have sufficient activity and strength. It will also show that they have the capacity to learn the additional duties and whether they have the proper degrees of courage as well as the ability to make critical decisions and reason under stress. There will be many officers, though promising in initial appearance and observation, who are found very unfit upon trial. Any member of the SWAT establishment found unfit for the duties should be rejected and replaced by better candidates.

It is not numbers, but bravery, intelligence, and skill that carry the day during a SWAT action.

After receiving their basic skills training, the operators and officers of the SWAT establishment should be trained on a regular schedule and be taught in the use of all SWAT arms by constant and routine exercise. This essential custom has never been emplaced in some agencies or has been allowed to lag behind or never been fully implemented. This simple custom has been abolished at times by the realization of long periods of few SWAT call-outs. It is not realistic to expect to find an operator or officer that is able to teach what they have never learned or experienced themselves. The only method for recovering and acquiring these customs and skill sets is by training and by consulting the older team members or through training with other agencies. Do not select or allow training that replicates only the exploits of past events. All training must challenge and maximize the ability of the team and its members to solve complicated problems. Training must also be dynamic so as to teach the current skills needed to resolve the life and death situations that may be faced in the future. The training must also realistically address present day realities as well as the public's expectations. Finally, use only those who are considered to be extremely respected and universally known to train the SWAT establishment.

Initial Operator Training

The first thing the operators must be taught is the proper use of their primary weapon, their handguns. A high level of competence can only be acquired by constant practice with practical realistic applications.

Nor is any skill of more consequence either in the dynamic search or slow, methodical search approaches. The operators must be comfortable in the use of their handguns and be able to use them with the greatest precision. Operators who do not use their primary firearm well are always in great danger of failing on a SWAT call. The operators should be able to complete departmental qualification courses with 100% accuracy, and no other level of qualification should be accepted. The operators must be able to account for every bullet fired with every weapon they use.

In particular, all of the operators should exercise frequently. This will allow them to move and capture fleeing suspects easily, take positions of advantage in deployment, and prevent suspects from escaping. When sent to reconnoiter and gather intelligence they must be able to advance with speed, return with great ease, and easily apprehend offenders when in pursuit.

Ascending and descending, traversing elevations and obstacles of all types, being able to climb up and down even the smallest of sheer surfaces, all while in full gear must be done without trouble or difficulty. There is a very material advantage to be derived from this type of exercise during a SWAT call-out, for an operator who advances with his firearm deployed, whom is able to run, climb, and move with celerity surprises offenders and puts the operator in position to resolve call-outs without force. These abilities take initiative away from offenders and places them in a positions of disadvantage should they attempt to strike at an operator. Without dispute the team must be physically fit and trained with all of its equipment. Only continual exercise in full gear accomplishes this goal.

Aerobic Conditioning

Without exception every operator should be in excellent physical condition, able to work in full gear for hours on even the hottest of days. It is sometimes impossible to resolve call-outs without traversing distances, and a foot pursuit of an offender may be necessary. A sudden danger may require the operators to change positions to contain an offender. The best SWAT teams will be perfect in every branch of their training. The training must also ensure cardio performance. Whenever deployed or in training, the operators should also be caring for their hydration needs. Allow them periodic breaks to help deal with the sweat and dust and to assist them in keeping fresh after their fatigues. The crisis negotiators and command post staff should also take care and make sure that they are also accustomed to the extremes they might face on a call-out, as they are also susceptible to these types of injuries.

Range Training

The past experience of range training shows that it must be done in full equipment and typically should use small targets. The best results are derived when physical and mental stress are also induced. The operators must train in full gear just as that worn during deployments. The training must be exacting, and the expectation of performance must be excellence. During training the operators should be using range targets smaller than those normally used by the agency. If the operators train in full gear, they will be familiar with all of their equipment and accustomed to carrying a full SWAT mission load. They should pass all qualification courses with 100% accuracy.

Training with full equipment is not only useful for the operators, but the crisis negotiators and command post team members must also be perfect with all of their assigned equipment. No team ever truly distinguished itself without having first trained under realistic conditions and performing full call-out training missions to confirm they were ready for a real mission. Every operator and officer, therefore, must perform all training at the range in full gear as if it were a real call-out and complete all qualifications and other training, whether using lethal or non-lethal force training, dynamic entry or slow, methodical search training. ICS and crisis negotiations should also train in full gear under realistic situations. Each operator and officer must be taught and be able to deploy in full gear; this will allow for the taking of every advantage against offenders. This will make them able to make arrests with a minimum of force and minimize the danger to the public while also giving the SWAT operators the greatest safety margin. The operators will perform as they train, so train them under the conditions they will face in the field and to the level you will expect them to perform at.

Shoot to Stop, Not to Wound

The SWAT operators will not be taught to wound an offender, nor should any officer be asked to disable limbs or shoot firearms from offender's hands. These types of scenarios are perpetuated by the media and the entertainment industry as being realistically possible. Seldom is this type of response a viable solution in true deadly force situations. Even though SWAT operators must attempt to use the least force necessary when making arrests, they must be trained precisely in the application of deadly force. Only lethal hits with firearms may be accepted during training. It must be remembered the purpose of any police use of force is to stop and arrest an offender. The operators will be shooting to stop offenders and generally the use of firearms is lethal. The use of lethal force must be precise because it exposes the public and other operators to lethal danger. All SWAT shootings must be conducted with the goal of striking bodily organs, only this will disable and stop an offender. This is the method regular police officers are taught and no greater expectation can be placed upon SWAT operators. This method will require a great amount of time and some expense for the operators to be competent. However, once acquired, the ability to use lethal force with absolute precision will give the team more options and in fact increase its ability to resolve any operation with a minimum loss of life.

Drilling by Scenario

All officers of the SWAT establishment must be trained regularly in the most realistic settings possible. It is this experience that yields the greatest training and advances in competency. It is important to use "Simunitions" type systems in training scenarios. Simunitions allow for realistic, and safe interactive training with firearms. The use of Simunitions affords certain proofs of the importance of cover and concealment as well as instilling the discipline of teamwork to resolve calls. This branch of drill is also effective for a department's regular officers as mistakes are instantly realized and corrected. These types of scenarios can be replayed until mastered, and those who do not succeed can instantly know why and are able to correct their mistakes.

All scenario training should allow for repetition until the operators know they can successfully resolve the situation and use their equipment perfectly prior to leaving the training site. Scenario training is never to be used as a measurement of "qualification." Qualifications are tests of the operators' skills. A failure to qualify on a test requires a return to training. Skill sets are acquired by perfect repetition.

Failure to train

No chief or sheriff can either be happy or secure if their department neglects the discipline of its SWAT establishment. A simple expenditure of money and purchase of equipment or the latest technological gadgets will not influence offenders or civil lawsuits, or keep the public's confidence in your team. This can only be affected by exact professionalism and the precise and necessary uses of force. Misconduct in the common affairs of policing may be retrieved, but that is quite otherwise on a SWAT call-out, where errors are compounded, sometimes fatally, and without remedy. These failures are always followed by intense media scrutiny and civil lawsuits. The consequences of engaging hostile offenders without skill or courage is that part of the public's confidence will be destroyed during a call-out gone wrong. The public will receive such an impression from the failure (defeat) that they may not trust their department's leadership again.

The Use of Long Arms

Besides the use of handguns, the operators must be furnished with long arms. These long arms typically consist of carbines and sub-machine guns. The operators should also be trained on the same small targets and again in full gear using realistic scenarios. The range masters must be careful to instruct the operators on how to properly load, reload, and unload these firearms as well as to aim them properly. They must be particularly trained on when to use deadly force. This practice increases accuracy and makes the operators excellent marksmen.

The Use of Ranged Intermediate Impact Weapons

All members of the SWAT team should be qualified as experts in the use of ranged intermediate impact weapons such as bean bag guns and forty-millimeter projectile launchers. The range masters who instruct on these weapons must be chosen with care. The operators must apply themselves diligently to properly use intermediate force and how to safely use intermediate force with lethal covering force. The operators must not intentionally strike lethal body areas when using intermediate-force weapons unless the situation calls for the use of lethal force. Any use of intermediate force must conform to the department's policy and procedure for applications of force. An operator can only acquire this skill with great application; again this skill can only be retained with exercise and practice.

The utility of ranged intermediate impact weapons in action has been demonstrated many times over on suicidal subject calls. The use of intermediate impact force to disable suicidal subjects long enough to be restrained from injuring themselves is frequently played out upon television by the media. The likelihood of success in dealing with suicidal subjects is greatly increased and looked upon favorably by the public when this type of method is employed. Having intermediate force weapons deployed gives the SWAT establishment the proper tools to implement the de-escalation of force and turns potential lethal force situations into intermediate force situations.

Precision-scoped Long Arms (Snipers)

Certain operators must have the ability to use precision-scoped firearms and observe offenders at a distance. These operators typically have a long history in the use of firearms and specialty training from the military. These operators must have the ability to continuously observe offenders for long periods of time and, if necessary, place a rifle round into the head from a distance of four hundred meters under extreme and stressful conditions. When employed, an observer and a sniper should always be paired together. These operators must be able to work as a team to use one bullet and successfully stop the offender. Snipers and their observers are the greatest means of ending hostage situations when negotiators are unable to induce the offender to surrender. The snipers and observers must be masters of concealment and hidden movement in order to place themselves in position to resolve a situation when needed.

Grenadiers

The use of grenadiers to resolve a situation is of the utmost importance. Deploying gas and other irritating agents blinds and irritates offenders. Irritants can also be used to deny an offender a certain area or move them from one area to another so they can be more effectively dealt with by your team. The use of chemical irritants gains initiative. The proper application of irritants requires great dexterity and skill as well as the ability to mathematically calculate the proper amounts of irritants to deploy into a structure. The use of chemical irritants to induce offenders to surrender is the preferred method for dealing with barricaded suspects. Irritants are introduced by hand grenade or in a ballistic projectile form. Additionally gas can be of the burning or non-burning types. "Burning gas" is a type of irritant in which the chemical is dispersed by a high temperature chemical release, this dispersal introduces the irritant into the atmosphere and has the potential to start flammable and combustibles on fire. "Non-burning" irritants are typically dispersed in projectile munitions that are simply released by breaking or otherwise opening the container and dispersing the contents over a larger area.

Keep and Use All Equipment Competently

The SWAT establishment must be well versed in all assigned equipment and have all of their equipment in good repair and instantly accessible to them. The SWAT establishment should carry the equipment home with them to be used at an instance's notice on a call-out if the agency does not possess the ability to deliver the equipment in a department vehicle. The SWAT team operators must have the ability to absolutely secure issued equipment at all times while being able to access and don their equipment at the slightest notice. The agency must have a suitable vehicle that is fully equipped to carry and deploy the SWAT gear and equipment the operators do not carry with them.

To Properly Maintain Issued Equipment and Be Ready for Instant Action (Call-out)

Operators must be accustomed to constantly care for and maintain all equipment. The entire establishment will be obliged to keep all personal equipment and all team equipment properly serviced and functional. The equipment must be ready at all times just as the entire SWAT establishment is ready. The entire SWAT establishment should carry provisions with them that would last for at least two days. Surplus MREs are perfect for this task. The use of pagers and phone trees to alert all members of call-outs is absolutely necessary. The entire SWAT establishment must become used to these practices. Failing to exercise diligence in this area will result in less-than-full team response.

Arms and Equipment

The SWAT operators personal equipment should consist of (at a minimum) a ballistic vest, helmet, goggles, radio, utility vest, suitable handgun holster, departmental handgun, handcuffs, flashlight, pepper spray or other irritant, uniform fatigues, and a less lethal option such as an expandable baton. The team and every operator must carry less lethal munitions and less lethal options on all call-outs as the goal is not to kill offenders but to arrest them. The use of extra body armor in the form of anti-ballistic greaves, side armor, neck armor, and ballistic face shields is useful and can be thought of as necessary, but not necessarily applicable to every SWAT call-out. The operators must still be able to move freely and perform all the functions of an arrest. The operators must also carry hydration with them as the weight and nature of their equipment and body armor will dehydrate them in nearly every situation.

The operators and all officers of the SWAT establishment must be able to respond with intermediate force on all operations, and all officers must carry handcuffs during SWAT actions. Frequently offenders respond in a fashion that requires an intermediate force response, and all must be prepared to accept a peaceful surrender. Carrying an intermediate-force option and handcuffs will allow all members of the SWAT establishment to respond appropriately and take the correct action to resolve the call-out.

Not all of your operators can be expected to carry ranged less-lethal impact weapons or irritants, but all should carry a baton and handcuffs. Nothing is worse than having a group of operators attempting to subdue an offender with bare hands, with all of the officers having to resort to piling on and tackling an offendor. This exposes their weapons and other equipment to possible use by the offender, not to mention the extra weight they must now cope with during an unnecessary wrestling match. The SWAT team may deploy "arrest teams." These are operators who are frequently deployed without long arms and are prepared to deal with non-compliant offenders whose actions do not rise to the level of lethal force. The operators assigned to make the physical arrest are the ones who should be without long arms but may also be deployed with ranged impact weapons and irritant gases in addition to their handguns. Carrying ranged impact weapons can also provide the SWAT team an opportunity to deal with vicious animals without destroying them or exposing the public to the hazards of gunfire.

The SWAT operators must wear ballistic gear including a helmet. There are numerous designs available. The helmet should be of Kevlar, and the operators must constantly train while wearing them to become accustomed to the use and weight of the helmet.

As to the long arms of the SWAT team, the standard .223 caliber military rifle or carbine is quite efficient. It penetrates most body armor available today and is accurate. Due to this round's ballistic performance, it typically fails to do much damage after penetrating walls and is thus safer for the general public. The long-range precision-scoped weapons operators should be using a rifle of the .308 caliber type or one of a similar ballistic capability. This weapon allows for a flat trajectory and increased precision in use. These calibers of rifles match that deployed by most policing agencies in the United States. Expert training in the use of these weapons types are also easily found and implemented.

Team Dispositions and Deployment

The SWAT team leaders should attempt to not initially deploy all of the operators in perimeter and entry positions on every call-out if possible. It is wise to keep a portion of the team in a position to rapidly respond to any new or unanticipated threats should an unexpected threat materialize or a sudden opportunity arise to resolve the situation. These operators should be placed in a position where they can become instantly mobile and yet still be rested. This reserve should not have outer perimeter duties if possible, but when a team is shorthanded, the reserve will generally come from perimeter positions. Team members placed in reserve need to be fully aware of the progress of the SWAT call at all times. In this manner, they can deploy and respond appropriately to any problem with great impetuosity. There have been many instances of reserve operators responding to crisis spots or capturing fleeing offenders.

The SWAT team will operate in the inner perimeter. Likewise the outer perimeter will typically be manned by officers controlled by the command post. The outer perimeter has the dual duty of helping to keep the suspect in while also keeping the public from entering into the danger of the inner perimeter. The outer perimeter can be manned by the regular officers of the department, but these officers must be alert to the hazards of their deployment and make sure they do not expose themselves to danger from the main offender while performing their duty. These members must also be mobile and have the ability to respond to a crisis area should and offender somehow become mobile.

The operators should be trained in all manners of shooting positions, and they should be able to use their handguns with both hands so as to be able to take advantage of all manner of cover and concealment. The operators should also be trained in using their handguns with just one hand in case one limb is disabled. This should include reloading and making a weapon either ready to use or safe. The operators must be able to perform these skills with either strong or weak hands.

Command Posts and Outer Perimeters

The position of the command post must be selected with care because improper selection of a command post location can compromise the successful resolution of a crisis. In a command post the incident command staff and crisis negotiators must be secure from the media. This will also allow the team leaders and supervisors to safely plan the team's response. Regular officers should be placed to guard entrances and even the perimeter of the command post from bystanders and the media. Neglecting this discipline will lead to erosion of public confidence and compromise the team's ability to respond to a call by exposing the plans of the team prior to their implementation.

A well-placed command post allows for briefing of the team without distractions. It allows for open discussion and the ability to find a successful tactical resolution to a crisis. The implications and ramifications of not having the media influence the decisions or reporting of the teams actions to an offender are extreme. The command post, especially in a rural residential area, must be chosen with great care. It should be easily secured by design, and there should be transportation access as well as access to water and restroom facilities if possible. The command post must not be exposed to the general public and should be secured from the media's prying eyes and ears.

The command post must have a separate staging area away from the operations planning area that has access to enough room to stage the multitude of emergency vehicles responding. The command post should be within distance of a media gathering spot and allow for a public information officer to have access to the overall scene commander. Remember, there will need to be separate areas within the command post for the crisis negotiators, incident command staff, and for the SWAT commanders and operators tactical command post. The command post may also be needed as a staging area for SWAT operators should the need arise.

There are two methods of establishing a command post. When the department's command post vehicle is not sufficiently sized, then an accessible building should be used. When the command post vehicle is used, care must be taken to site the van away from danger of the call-out itself. This allows command staff to function without fear of being involved in the call-out or having to take direct action themselves. The command post should be outside the outer perimeter of the call, but close enough to immediately deploy officers to the outer perimeter and allow for easy deployment of operators into the inner perimeter.

The command post should be free from exterior distractions such as traffic noise from passing vehicles or other heavy equipment. Should a building be utilized as a command post, the building must have complete access for communications and restroom facilities. If possible, proper heating and cooling should also be sought out. It is extremely important that the proper area be selected for a command post. The crisis negotiators must be able to operate without interference or interruption. The ICS team must also be able to secure the supplies, equipment, and make the plans necessary to help the SWAT operators resolve the call safely with a minimum of risk to or hindrance from the public.

Evolutions and team movements

No part of drill is more essential than for the operators continually using cover and concealment while operating as a team to counter all threats. The greatest exactness should be used whenever possible in slow and methodical searching. Moving too fast unnecessarily exposes operators to hazards and lessens their accuracy with firearms. If the operators move too fast it increases the odds of detection by an offender. It also does not allow the operators to take the greatest advantage of cover and concealment to gain positions of advantage. Even though most searching for offenders is done using the slow and methodical type of search method, the operators must constantly be trained in both dynamic entry and slow and methodical entry techniques. The team must be able to evacuate at a moment notice, make sudden arrests, shift positions, and make an assault at a moments notice. The team must be able to conduct these types of maneuvers in urban and rural areas, as well as in open country. During slow and methodical movement, the team should be able to function in skirmish lines and be able to use basic hand signals to communicate silently while in formation and maintaining covering positions. The operators must be taught to search all of the types of terrain in the jurisdiction served, and they must be able to do so while using the various other special units available in the department, such as canines.

Monthly Training

It is the policy of the great SWAT teams to train regularly. In California the teams must train in accordance with their roles (by statute). Full-time teams must train at least twenty-four hours a month. Part-time or auxiliary-duty teams must train at least sixteen hours a month. It should be noted that these are absolute minimums and this training is just to maintain basic skills. When new policing skills and methods are introduced to the teams then additional training time should be allotted. Additionally, operators should be sent to specialized schools in order to obtain the latest training and skills sets being used and accepted as expert.

The crisis negotiators and incident command officers as well should be exercised in their crafts with scenarios presented that require them to make decisions and go through the evolutions necessary to acquire equipment and logistics and facilitate the surrender of suspects. All training for all parts of the SWAT establishment should incorporate on-going, realistic, scenario-based training that requires the operators and officers to make decisions under realistic stressful conditions. The scenarios should be repeated until the operators and officers succeed and should prepare the team for all of the various situations they will be required to respond to.

Conclusion

These SWAT maxims and instructions are proof of my devotion and zeal for SWAT service. I have carefully collected them from the works of all the proven departments open on the subject. My design herein is to point out the certain method of forming good and serviceable SWAT teams. This can only be accomplished by an exact imitation of the successful SWAT teams of today and in the care in the choice and discipline of operators and officers for the SWAT establishment.

Every department has of officers of courage. Every department also has the ability to produce an excellent SWAT team or successful SWAT team member. Look to the leaders in SWAT and follow their examples. The Los Angeles Police Department has acquired the greatest reputation in SWAT because of their constant dedication to perfection in training and advancement of the art of SWAT. The Los Angeles Sheriff's Department has become so proficient in SWAT as to be asked to train with DELTA, SAS, and other international police and military organizations. It is well known that there are numerous other departments throughout the United States that have excellent SWAT establishments.

There is no need to enumerate the different SWAT establishments across our nation and detail how extremely efficient and formidable they are. However, we must realize they did not become excellent in their delivery of SWAT service in any way except through rigorous training and attention to discipline. Superintendents and command staff must be aware of the false sense of security that may be established by long periods of no call-outs. These false feelings of security can alter some departments' dispositions, draw resources from SWAT to other specialty units, and infuse into its agency a feeling that SWAT will not be needed, or the agency no longer needs to maintain a high level of SWAT excellence. Without care, the realization of SWAT discipline is first neglected, falters, and then sinks to oblivion.

It should not be surprising that this alteration happens even now after all of the growing concern over violent gangs and terrorism. Few consider anything but the relatively peaceful times they might have enjoyed since their last truly difficult crisis. As a departmental superintendent you must be cognizant and alerted to the increasing threats of today's world. You need to consider that your department can never fail on a SWAT mission. Allowing idleness and neglect of discipline will ensure that the next time your SWAT teams may not be able to stand between a violent offender. This does not even consider the potential for a possible confrontation and possibly avoiding the devastating effects of a terrorist attack. Remember, the offender need only be lucky once; your SWAT team must be very, very good *every* time. You must be convinced that discipline cannot be too often inculcated or practiced, and strict attention to the choice of and training your SWAT establishment must also be undertaken. It is also certain that it is ultimately much less expensive for a department to use its SWAT operators, rather than having to pay the price for using and relying upon specialty units or department teams that attempt to handle situations they are neither trained or equipped to handle.

2

ORGANIZATION OF THE SWAT ESTABLISHMENT

Preface to Chapter 2—To My Sheriff/Chief

The fact that your department has successfully resolved major critical incidents in the past does not prove incontestably that your department has a full and perfect knowledge of the SWAT discipline. Success in the resolution of critical incidents is not the most certain mark of skill. You have risen to the position of a superintendent because of your wisdom and comprehension of the duties and responsibilities of the agency are far greater than others. You have also most likely sought instruction and education from the best experts and training available in this field, not withstanding your department's own successful exploits and resolution of critical incidents.

It is my hope this treatise is not used as policy and procedure, but more as a reference or guideline. Its purpose is not to impose rules for a superintendent to obey. The superintendent is in control and responsible for all departmental actions. Disobedience to the will of a superintendent would be insubordination and could be both criminally and civilly dangerous. The purpose of these works are to encourage obedience and pass my knowledge and experiences in this field. The purpose of the first chapter was to impart knowledge on the proper choice and discipline of new operators. The next chapter's purpose is to explain the roles and organization of the SWAT establishment. This will also assist in command and control during operations.

Incident Command Systems and Crisis Negotiators

The SWAT establishment consists of three parts: the SWAT team, crisis negotiations, and incident command.

The crisis negotiators are so called for their ability to defuse crisis through intermediation. They intercede between the trigger of the event and offer a peaceful resolution to the offender. The crisis negotiators are an important and critical portion of every SWAT team and are organized and equipped for communication with offenders. Crisis negotiators are responsible for the passive imposition of the SWAT establishment team commander's will upon offenders.

Incident command team (ICS) consists of two divisions, one of officers selected to maintain the outer perimeter, and the other composed of officers and supervisors who ensure the material needs of each call-out are met. The ICS is employed to provision the SWAT establishment for the entire event, securing liaison and access to other public agencies that may be needed to solve the crisis, and to also protect the general public from the factors of the event.

The SWAT team is actually deployed to resolve the call; they are positioned in the inner perimeter and keep the threat contained within until resolved. SWAT officers should not be used if possible to garrison outer perimeters. The SWAT operators are certainly the most versatile and are certainly the most useful and necessary officers the agency has. The SWAT team itself is generally divided into two corps—the entry team and the perimeter team. The operators assigned to long-range precision weapons (sometimes called "scouts" or "snipers") train mostly with the perimeter team and away from the entry team. The peculiar strength of the SWAT team always consists in the excellent organization of their SWAT establishment. These strengths are dominated by the care and exactness used in the choice of its officers. The number of SWAT operators in a department is generally much more considerable than that of the crisis negotiators and the ICS teams.

Differences Between SWAT Operators and Support Staff

The larger departments form their teams into full-time units and typically field squads of SWAT operators at twelve to sixteen apiece. The smaller departments typically have part-time teams and assignment to SWAT is as a secondary or ancillary position. There is a big difference in the duties of a SWAT operator and the rest of the support staff of the SWAT establishment. The latter are assigned to crisis negotiators or ICS. Their main duties are to help induce a surrender, to keep the SWAT team supplied with intelligence, equipments, and supplies, and to keep the public from being injured or otherwise becoming involved in the incident.

In some departments the SWAT operators and support staff have little knowledge of one another or any ties of affection. Each of these units has its own peculiar disciplines, customs, and manner of operations. Little can be expected from officers assigned so dissimilar roles in resolution of a crisis if they do not train together. It is one of the most essential points in SWAT undertakings that the entire establishment should be put in motion and governed by one and the same order. It is almost impossible for the officers and operators to act in concert under such varying and unsettled circumstances if they have not practiced together and do not understand the roles of their counterparts. They are, however, when properly trained and disciplined, of material service and are always joined as one team when on an operation. Although the SWAT team operators typically do not place their principal dependence on support staff, integration of the entire SWAT establishment is critical to the successful resolution of every call.

The complete SWAT establishment, in its own a peculiar team, contains within itself the heavily-armed entry personnel: the breachers, gas grenadiers, less lethal specialists, scouts, and snipers, the medical technicians, together with the special equipment operators incorporated within it. These bodies are all motivated with the same spirit and are united inseparably in their various dispositions for forming, securing, isolating, and arresting a suspect. The SWAT team should be compact and perfect in all its parts and, without any foreign assistance, have always been superior to any patrol team or other specialty unit to compare it with. The SWAT team's greatness is a proof of the excellence of their officers. A SWAT team must successfully resolve all situations they are assigned or are forced to engage in.

A Name Does Not Make or Define a Team

There are many different names for the SWAT establishment, "ERT" (Emergency Response Team) and "SRT" (Special Response Team) being just a few of the names taken up by various departments to label and define their crisis response teams. No matter the name of the unit, it is the role its team members take on that defines whether or not it is truly part of the SWAT establishment. The duties of being assigned specifically to resolve dangerous crisis situations through specialized tactics means being a part of the SWAT establishment. In many instances groups of team members are in fact not a SWAT team even though they may have formal titles that hint they are assigned to the duties of the SWAT establishment. Training for and regularly handling high risk crisis situations defines those truly assigned to the SWAT establishment.

Common Causes of Decay of a SWAT Establishment

Due to neglect by predecessors, unearned honors, and preferments, which were formerly the recompenses of merit and long service, positions in specialized units are sometimes staffed out of interest and favor. The SWAT establishment is not brought into being by the introduction or changing of a unit's formal title or by merely assigning officers to it. Hard work and discipline through training define the team. Care is sometimes not exercised to properly replace or retire injured officers who after serving their time should be discharged from SWAT service. Team vacancies will continually happen due to injury, promotion, and other factors. If a team is not supplied every year or does not have a system set in place to replace the operators immediately on an as needed basis, in time the team will become disabled. What once may have been an effective team can be easily reduced to inefficiency through simple neglect. Another cause of weakness in teams is that in some instances team leaders and supervisors are improperly appointed the duty. The equipment load is too heavy for them, or they do not hold or even understand the need for SWAT disciplines.

To avoid this type of inconvenience, those with prior service in the SWAT establishment should be the most strongly considered for this service, especially those actually on the SWAT team as they will need to also function at times as an operator. Those with prior service experience will better understand and meet the physical requirements of the assignment. They will also understand their need to perform as well as how to have the SWAT team perform at a high operational level.

The brave consequences of the SWAT team are rarely if ever recognized. The SWAT operators will regularly undertake far more risk than the common officers and detectives of the agency. Remember when it is time to promote officers to supervisory positions that nothing is so equal as to be judged equally. True experience can only be gained "under fire" by resolving real life crisis situations under stress. Those who can make effective decisions under this type of stress will continue to make good decisions in other roles they may be trained for. There is only one true way to get experience in resolving high-risk crisis situations, training for and handling them.

The expense of keeping up the SWAT establishment and keeping the SWAT establishment at a high operational level is the responsibility of the department command staff and superintendent. The result of purchasing inferior equipment, not replacing broken or maintaining worn equipment, and not securing necessary training is that the SWAT establishment will deploy without the necessary tools and disciplines to safely resolve the crisis. The costs of recovering or achieving the excellence in discipline lost through inattention and neglect will be compounded exponentially by a civil lawsuit. It is cheaper to replace the equipment than to pay off a legal tort.

The best SWAT establishments continually look for and correct abuses or lack of attention to training and equipment in addition to continually looking for more efficient ways to resolve crisis situations. This reformation leads to advantages that allow the SWAT establishment to succeed. The department's posterity and the community's safety in critical events depend very much upon dedicating the proper resources to the SWAT establishment.

Organization of the SWAT Establishment

There is considerable difference within the SWAT community on the perfect size of a SWAT team. What is known is that the team must be sufficient for the task at hand, and they must practice economy of force. Do not constantly rely upon sending a few good officers to accomplish a mission or supplement a specialty unit with SWAT operators to resolve a SWAT mission. If this description appears imperfect it is not imputed to me, but to the difficulty and varying complexity of the crisis situations a department will be tasked with resolving. A department superintendent skilled in law enforcement administration can make themselves and the agency they serve invincible to public criticism by maintaining a properly trained, equipped, and disciplined SWAT establishment.

The SWAT team must be carefully chosen with particular attention paid to their character and physical attributes. They will be continually and realistically trained at least twice a month, and the entire SWAT establishment must be led and commanded by supervisors personally selected by command staff. The department and the chief superintendent's reputation will rely upon and be imprinted on the community by the SWAT establishments performance. The operators will risk their lives and futures; remember this and give them the love and veneration they deserve. Once the title of SWAT commander is bestowed, subordinates are bound to give the most sincere devotion and homage. The SWAT establishment will represent its commander as well as the department's chief superintendent and command staff in the most public of situations.

All of the officers in the SWAT establishment serve faithfully and under the SWAT commander's authority. The operators and officers selected for your SWAT establishment will willingly and implicitly impose the SWAT commanders will in crisis situations; they will obey every command and place themselves in positions of the greatest danger to interpose themselves between the community and the danger! The actual SWAT establishment should consist of an entry team, a perimeter team, crisis negotiators, and incident command staff.

The SWAT team will consist of entry team, breachers, perimeter team, explosives experts, less lethal experts, scouts, snipers, and grenadiers. The crisis negotiation team consists of primary negotiators, back ups, coaches, supervisors, and intelligence gatherers. These team members are usually posted away from the inner perimeter very close to your command post. This allows them to give their SWAT commander the latest and best information available on the offender's state of mind with which to make decisions that allow for the to safe arrest of offenders.

The third part of the SWAT establishment is made up of the incident command/emergency management team (ICS). The ICS team and its commander are responsible for the overall resolution of the incident. No SWAT team should ever be deployed without consideration of a full ICS call-out. The ICS team will keep the public from danger, help with outer perimeter containment, and ensure the logistical success of the operation. Failing to feed and or equip the SWAT establishment with proper tools and supplies will keep them from safely resolving calls. Failure to pay proper attention to all three parts of the SWAT establishment will ultimately ensure a failure during a crisis.

SWAT Operators

The entry team members are primarily trained in dealing with offenders inside of structures or confined areas and any contingencies occurring inside the inner perimeter. The perimeter team is primarily trained and assigned to contain the incident within the inner perimeter and function as entry team members in emergencies. These operators frequently become the arrest team and take an offender into custody, and then escort the offender out to the command post. Scouts (consisting of observers and snipers, or long-range precision rifle teams) are used to gather intelligence, cover the approach of entry team members and perimeter team members, and position themselves to use their long-range precision weapons if necessary.

Breachers will typically be part of the entry team and train with them. Breachers will use mechanical devices and explosives to force entry into structures. Breachers may also have training in lock-smithing techniques for defeating locking mechanisms. Grenadiers will be used to introduce chemical agents and irritants to offenders and can be assigned the use of ranged intermediate impact weapons. Grenadiers will usually be attached to perimeter teams and are able to introduce chemical agents (usually irritants) from a distance with projectile launchers for safety. However, the deployment of gas can be done by any operator, and all operators should be trained in deploying irritants.

Crisis negotiators will make contact with offenders and provide the best opportunity for creating a peaceful resolution to a crisis. Crisis negotiators operate by exchanging small considerations and bargain for the release of any hostages and eventual surrender of offenders. Crisis negotiators also give SWAT commanders critical information on the mindset of offenders and any criminal intelligence they have gained so that informed tactical decisions can be planned by the

SWAT team. Crisis negotiators may need the SWAT team itself to deploy communications devices to the offender such as throw phones or cell phones to actually make contact with the offender.

Incident command staff typically man the outer perimeter and must be mindful to keep the public from being injured as well as keeping the news media from wandering into the crime event. Just as importantly, they are responsible for securing the necessary supplies and logistics needed to resolve the situation. They keep the SWAT establishment in the field fully supplied with the equipment and supplies needed to resolve the situation.

Command Staff Duties

Command staff of the SWAT establishment should not be sent on operations with the SWAT team if at all possible. Their authority is extended and exercised in most situations by radio. SWAT operators should be led by the team sergeants, corporals, and other team leaders in the field. The peculiar position of a team commander inside the perimeter should be avoided whenever possible as the commander may not be able to detach themselves far enough from the critical events to make decisions. The decisions of the team commander are to guide the sergeants and team members, to grant authority for a lawful and necessary use of force or other action to resolve situation.

The team commander has charge of all arms, equipment, and provisions. It is also the team commander's duty to maintain contact with the incident command staff and crisis negotiations supervisors. The

SWAT commander ought to be a careful and intelligent individual as he has the sole charge of training and deploying the SWAT establishment. They must also be able to extract obedience and excellence from the SWAT establishment and thus effectively impose their will upon the offender. Successful resolution of the crisis by the team is entirely to the SWAT commander's honor and credit.

The incident command staff is in charge of the procurement of supplies and equipment, as well as having the important duty of maintaining a secure perimeter and protecting the command post. This duty is of no small importance. The position of the command post, the direction of support, the inspection of the outer perimeter, and the actual consumption of resources depend upon the ICS commander. The ICS commander's authority extends over the evacuation of the wounded from and the deployment of medical personnel to the outer perimeter, provides for ambulances, any transportation for the any SWAT team, and acquiring necessary tools not normally carried by the SWAT team.

The ICS commander is likewise responsible for replenishing expended munitions and food consumed by the operation. This commander will secure air support and other instruments or large pieces of equipment from other government agencies as well as calling in additional SWAT teams from other agencies for extended or extraordinary events. This post always requires an officer of great skill, experience, and long service, who has consequently proved capable of instructing others in this branch of the profession.

The Outer Perimeter

The outer perimeter will most likely be composed of regular patrol officers. These officers will be responsible for the management of the crime scene as well as keeping the public out of the event and the media at bay. A proficient supervisor must be put in charge of this position and must constantly attend to the perimeter's duties. The perimeter officers may be needed should offenders somehow manage to go mobile or otherwise escape the inner perimeter. More than likely these officers will be used to block streets and thoroughfares to protect the public. The officers of the outer perimeter should also be alert for opportunities to assist in gathering intelligence on offenders for use by the SWAT team.

Supervisors of the Operators

The actual SWAT team supervisors should be operators of experience themselves. They must be educated in their field and have excellent reputations within the department. They must be at a minimum technically proficient as operators and they must have excellent knowledge in the disciplines and applications of SWAT tactics. They should be distinguished for their own skill in SWAT exercises and evolutions, for the advantage of their person and integrity of their ability. A SWAT supervisor is an integral team member and may have to perform the duties of a line operator during a mission. For this reason their discipline must be so strict as to be exercised to at least the same level as the operators. The SWAT supervisors should be so perfect in the SWAT exercises as to set the example and be able lead from the front when necessary. SWAT supervisors must realize that their own abilities are reflected in the appearance and discipline of their operators. SWAT suupervisors should be evaluated on how their team performs. The entire team must be neat and clean, their equipment well cared for and properly worn, and the utmost attention should be paid by the supervisor to the SWAT team's actual performance in resolving crises.

Swat Team Divisions

The SWAT team should be divided into sections, and the team members should train for duties in all SWAT disciplines with their appointed supervisors and team leaders. In this manner, operators will be able to perform perimeter or entry duties as necessary and have the ability to perform in both roles. Additionally the operators will become used to their supervisor's methods of operation, as well as know the expectations for their performance. This will also allow the supervisor to assess each operator's progress and performance during training and on each call-out. It will allow for proper deployment and later a critical review that yields positive results. The result will be greater unit cohesion and constant improvement of performance.

Crisis Negotiators

As the SWAT team is divided up, so should the crisis negotiations team. The primary negotiators make contact with the offender, while a secondary negotiator gets ready to replace a primary negotiator if needed. A coach helps the primary negotiator and disconnects from the actual emotions of the negotiation. Finally the negotiations team leader relays intelligence gathered from the offender and informs the SWAT commander and the rest of the SWAT establishment (via the SWAT commander) of the offender's state of mind and likely or ongoing actions.

Drawing Up the SWAT Establishment for a Call-out

It is important to know of the basic drawing up of the SWAT establishment's deployment on a call-out. In most instances the general deployment of the team will be greatly influenced by the extraordinary circumstances of each call-out, thus there is no set deployment. Generally, the entry team is posted in positions of advantage inside the inner perimeter. The perimeter team is posted at and creates the inner perimeter. The scout snipers deploy in positions of advantage and may even be deployed far outside of the outer perimeter. The incident command post is deployed outside of the outer perimeter. The crisis negotiations team is deployed in or near the command post.

The entry team shall generally be composed of lead operators or "scouts," entry operators, and breachers. These operators may be armed with sub-machine guns or assault rifles in addition to their pistols. All will wear full body armor and some may be tasked with carrying ballistic shields and breaching equipment. Their weapons may include less lethal ranged impact weapons, gas and distraction devices, as well as common mechanical and explosive breaching tools. These operators will be particularly exercised in the use of their long arms and sub-machine guns they carry; they will be expected to be able to deploy them with three round burst or fully automatic potential equally efficiently.

The standard shooting response for all officers is two to three rounds whether deploying pistols, sub-guns, or .223-caliber rounds. This allows for stopping power as the human body has the ability to shut down or significantly slow bleeding to single injury areas but cannot stop bleeding to multiple injuries.

The inner perimeter team will be deployed with assault rifles and their handguns. They will also be accompanied by grenadiers and ranged less-lethal impact weapons. They may or may not be deployed in full body armor. Any police K9s attached to the SWAT establishment will usually begin deployment in the inner perimeter. The K9s and other perimeter officers may become part of the entry team if needed and must be trained to do so. The perimeter team's main function is to keep a close containment of offender and gather intelligence to help resolve the event. K-9's are frequently used to search for and locate suspects and to minimize the danger to operators during these times.

The crisis negotiators will usually be deployed next to or in a secure position within the command post. They will be establishing and maintaining contact with offenders. Their duties are to convince and bargain the offenders into a surrender. They will also attempt to secure the release of any hostages and to gather intelligence on the motives and possible actions that will be taken by the offender. It is of the utmost importance that the primary negotiators utilize the coaches and provide the crisis negotiations supervisor with current updates at all times. There is a tendency for primary negotiators to become emotionally attached or fail to see all facets and ramifications of their contact with offenders. The supervisor and coach must remain detached from contact with the offender so as to make completely informed decisions and pass valid intelligence onto the SWAT commander so that informed tactical decisions can be made to resolve the call. The task of primary negotiation is difficult and requires complete concentration on the tactics and strategy of human communications.

The ICS team helps to plan and initiate the outlay of resources and secures items of equipment or other supplies necessary to fully resolve the incident. The ICS team also is responsible for security of the outer perimeter, keeping the public from danger, functioning as a reserve for the inner perimeter, preparing the turning of the crime scene over to detectives, and maintaining the security and smooth flow of operations within the command post and crisis negotiations teams.

Communications

Special care must be taken in communications within the SWAT establishment and with other agencies. The confusion and subsequent friction caused by an active operation means the dedication of a dispatcher and recorder is imperative to any operation. Each operator must have a multi-channel radio to facilitate multiple conversations between the SWAT establishment and the rest of the department, as well as teams and officers from other departments. Furthermore all the SWAT operators should all be easily distinguishable from other officers. Their dress should be of a uniform that allows for camouflage in urban and rural areas of the jurisdiction served.

It should be noted that the confidence of the public is absolutely necessary and the goal of the SWAT establishment is to resolve events. While doing so they must not overly intimidate or frighten the public as this erodes confidence and can make the solving of the event worse than acts of the event that caused the call-out. The entire SWAT establishment must remember this in their operations and conduct therein, their duty is not to intimidate the public they serve, their job is to make the public feel safe and intimate the offenders.

Being Held Publicly Accountable

The keeping of logistics requires officers who have been properly trained. The superintendents should select worthy candidates who keep account of the cost of an operation to make sure the financial abilities of the department are not ruined and the municipality served is not broken by the expenditures. The cost of the call-out will be enormous; the cost of not responding in full or successfully resolving a call-out is even higher due to civil litigation. Do not allow the SWAT establishment to be short changed when they are being sent into danger. The daily costs and training costs will also be high; again, the civil liability incurred by not being fully trained and documented as having been trained are even higher. Civil lawsuits will result in the loss of agency funding from a disaffected public or the municipality's elected officials. The loss of confidence within the department over a failed SWAT operation will be compounded by the loss of confidence within the community served.

Every decision the SWAT establishment makes in tactics or training, selection of equipment, selection of officers and operators, or any decision undertaken to resolve a crisis will ultimately need to stand up to public scrutiny. The SWAT establishment is responsible and must answer to the agency superintendents. The superintendents must answer to the public they serve.

Holding the SWAT Establishment Accountable

The entire SWAT establishment must be prepared to be on duty and available for a call-out response at all times. The establishment must be ready for duty twenty-four hours a day, seven days a week, 365 days a year. This may require them to take vital pieces of equipment home with them, such as body armor and personal weapons. They must have sufficient vehicles for deployment and transport of supplies and equipment available at all times. They should receive additional pay as this the only fair remedy for the requirement and price exacted from the SWAT operators and the entire SWAT establishment. Finally, the SWAT establishment must be able to account for all of the equipment, supplies, and man hours expended at all times.

Promotion within the SWAT Establishment

It is known that good police officers do not necessarily make good police supervisors. Alternatively, it is known that lousy police officers make lousy police supervisors. If the physical rigors of being a SWAT operator are added into this equation, it is easy to see the need to never dismiss a capable former operator from consideration of promotion to a supervisory position within the SWAT establishment. The SWAT establishment supervisors cannot be dismissed from any of the assignments their officers or operators may have to perform as they will be expected to perform them in an emergency or by design. The SWAT supervisors must carry the same weights of equipment. They must also be able to operate the same equipment and function in the role of an operator.

The best SWAT establishment supervisors will thus be promoted from within the establishment itself as they understand and have the experience and requirements of training and decision making under stress derived from actual service. Rotation of SWAT supervisors to other duties in exchange for police supervisors with no SWAT experience is highly discouraged. Certainly, some extremely bright and talented supervisors may need to be selected and trained for a SWAT supervisory position. If this is a necessity for the agency, extreme care and close mentoring by other highly qualified SWAT supervisors and team leaders must be implemented.

SWAT Communications

The use of proper communications is of extreme importance due to the confusing nature and general bedlam that occurs during active calls. The SWAT operators must be able to communicate using hand signals, verbally, as well as by radio. This means that the SWAT operators and crisis negotiators must have the ability to speak in the languages spoken in the community served. If a particular language is being spoken at a crisis then at a minimum the establishment must have the ability to locate a translator and have the ability to inject the translator into the crisis.

Drills and Training

With the organization of the SWAT establishment explained the next step is to return to training. The newer operators must be sent to a basic operator's school. This school will usually consist of eighty hours of basic SWAT training. After this training, the operator should be able to join the team and function with the team in team training. However, this training is a bare minimum and at least sixteen hours (or more) a month is required to maintain the minimum amount of expertise for the operators. New disciplines and skill sets such as repelling, grenadier, breaching, and use of other special weapons require more training, generally derived in forty-hour blocks. Only through practice will true mastery of the various tools, skill sets, and weapons used will the operators become completely proficient.

In the times of utmost stress, the operators and

officers will resort to their last training. Training must be as realistic and demanding as possible. If the operators and officers have not trained for a situation, the result of the next crisis will not be known before hand. Untrained operators and officers generally fail in some manner; sooner or later that failure will turn fatal to the operator, the offender, or the public they serve.

The operators are to be exercised in full gear in conditions as near as possible to those they will encounter in the field. They should be wearing their body armor and regularly train in gaseous (irritant) atmospheres. They should be able to perform all aspects of an arrest and perform all manners of the department's use of force spectrum while in full gear. They must be able to advance and retire, hold positions for long periods of time, and be especially adept at the use of cover and concealment. They must be disciplined in their use of force. They must be excellent marksman and be able to service their weapons in every manner.

The SWAT team's qualification courses should mirror or strongly resemble those of the department. Additionally qualification should not be counted as training. Qualification is proof of training. All operators should be qualified in the use of all means of delivery of the use of force approved by the department. This will require operators to qualify in full deployment gear with baton, chemical irritants, precision weapons, sub-guns, pistols, assault rifles, shotguns, ranged impact weapons, and the use of all bodily force options.

The SWAT operators must train without interruption at all times of the year as they must be able to function at all times of the year. In short, if it is raining, they must train in the rain; when it is one hundred degrees, they must train in the heat. The operator must also be able to function in every part of the jurisdiction served. If there are rivers or other bodies of water, they must be able to function in them; if there is desert or mountainous areas, they must be able to thrive in those areas as well. On operations discipline is the superior strength. If there is no discipline there is no team. The very essence of the establishment and successful operations resolution is that they must constantly practice the art of SWAT to successfully implement it.

Machines and Tools of SWAT

The SWAT team owes its success to the use of its arms and equipment as well as to the courage and skill of its operators. There should be an armored vehicle capable of withstanding rifle fire to transport the operators into danger areas. These armored vehicles may range from ex-military transports to the armored cars similar to those used by banking institutions to transport money. This equipment will be used not only to transport the operators into and out of dangerous areas, but also to evacuate wounded victims or innocent bystanders from danger. An armored vehicle can also emplace or provide moveable cover and is proof against small arms fire. Besides the armored vehicle, the jurisdiction may also need to use helicopters for deployment of operators into otherwise inaccessible places or places in which instance access is necessary. The proper procedures for boarding and debarking must be practiced over and over in full gear by your operators. Helicopters may also perform reconnaissance and provide intelligence to help resolve the crisis.

The SWAT team should also carry with it ropes and

rappel gear, as well as ballistic shields and mechanical robots, picks, rams, ladders, pole cameras, throw phones, and other devices to insert gases and distraction devices. The ropes and rappel gear help provide the ability to ascend and descend heights. Ballistic shields provide proof against handgun fire only.

Mechanical robots may provide a safe reconnaissance or the ability to detect or diffuse booby traps. Picks and rams are used to breach or force entry into a structure. Pole cameras are used to safely view offenders or otherwise safely access areas prior to entry by the SWAT team. Throw phones are used by the SWAT team to assist the crisis negotiators in communicating with offenders. Throw phones may also contain video cameras that can be used to gain intelligence of the offenders activities.

3

DISPOSITIONS FOR ACTION

Preface to Chapter 3—To My Sheriff/Chief

The Los Angeles Police Department and Los Angeles Sheriff's Department were masters of SWAT before all others. They both have excelled not only in the art of SWAT but also in other arts and sciences of policing. The Los Angeles Police Department was the first agency in the United States to create a SWAT team. They are also affirmed to be the first who have reasoned on the events of SWAT actions and have committed their observations thereon to writing with such success as to induce other major and minor agencies to follow their examples. They do not consider being totally dependent on the courage and ability of the average line officers to handle emergency crisis; they have and yield to certain rules and fixed principles to resolve violent crisis. As a consequence they are recognized as experts and now teach in established schools the tactics for the instruction of the art of SWAT maneuvers and disciplines.

Both of these agencies are truly worthy of admiration and have particularly applied themselves to the study of this art, without which policing would ultimately fail. All other agencies have followed their example, and have practiced their SWAT establishments those lessons imparted, other agencies believe so strongly in these methods as to make their disciplines policy and procedure. Their maxims and instructions have been dispersed through the works of the California Police Officer Research Association, the California Police Officers Standardized Training, the California Tactical Officers Association, and even National Tactical Officers Association.

The sum of their works would be too tedious to include in this short book, and my authority and ability to recount their proficiency would be unsatisfactory. The effect of the Los Angeles Police Department and Los Angeles Sheriff's Department's skill in dispositions for SWAT actions appears most prominently in the single instance of the Hollywood shootout, where the Los Angeles Police Department's SWAT team assisted patrol officers in arresting extremely heavily armed and armored violent offenders who were using automatic weapons to fend off numerous patrol officers. It was not merely superior skill and conduct that defeated those offenders and ended the crisis, it was the team's discipline and valor, knowing what they must and had to do to bring the crisis to a successful conclusion that was being nationally televised. Had the SWAT operators waited for formal permission or been chained to a tedious command structure to practice their skills the offenders would have been responsible for the deaths of numerous police officers and innocent bystanders. Only the direct SWAT intervention saved so many lives.

If anyone believes that the same type of violent crisis that needs to be immediately addressed will never occur in their jurisdiction, simply contact the residents near the sleepy neighborhood of Columbine High School in Colorado or any of the students and instructors of peaceful West Virginia University and compare similarities. Those, therefore, who desire peaceful policing solutions, should prepare for the use of force to keep that peace. Those who aspire to successfully resolve those violent crisis, should spare no pains to form their SWAT establishment, and those who hope for continued SWAT successes, should operate on principle, not chance. No one dares to attack or insult a power of known superiority in SWAT actions.

The Numbers Which Should Compose the SWAT Establishment

The first book entreats the choice and exercises of new operators and officers. The second book explains the establishment and methods of discipline, and the third book contains the dispositions for SWAT actions. By this methodical progression in the following of instruction on general actions a method to successful resolution of crisis calls will be better understood and of greater use. The term *SWAT establishment* refers to the number of operators, crisis negotiators, and ICS personnel. The establishments that fail to successfully resolve call-outs typically do not do so for want of numbers; their failures are almost always due to their lack of discipline and training. If the team is too big, then it will not trained sufficiently with the resources available.

Failure to train all members sufficiently is planning to fail. Call-outs are often dynamic and changing, and there is no guarantee of human behavior that lets the commander and team supervisors safely ascertain the actions of every offender. This means that inevitably the lowest trained operator will be in a position of extreme importance, making critical decisions and having to take action to a situation they are not fully prepared or trained to handle. Additionally operators are likely to use each piece of equipment issued, if not fully trained the operator will fail because of inability to operate that equipment under stress.

The establishment must be large enough to respond to the amount of calls for service the jurisdiction normally responds too. The likelihood of having multiple SWAT call-outs must also be considered, the larger the jurisdiction the more likelihood this will occur. Finally the agency should be prepared and have in place working agreements for mutual aid with other nearby agencies. Having a mutual aid agreement is one thing; having a working mutual aid agreement is another. This means that the SWAT establishments must train with each other under realistic circumstances and all aspects of the SWAT call-outs must be addressed. Special attention must be paid and prior intra-agency training conducted to ensure a smooth transition during a handoff of duties during an operation, as well as ensuring that all members of the SWAT establishment are able to work with members of assisting teams. Policy and procedure as well as chain of command and communications issues must all be resolved whenever possible prior to any joint call-outs.

When to Call-out

The department should have documentation in the form of policy and procedure on what criteria warrants a SWAT call-out. Many agencies use specialty units and detective units to serve common search warrants. This practice is necessary. What is lacking many times is an established procedure to distinguish what a specialty unit should be handling versus what the SWAT team should be handling.

Means of Preserving Health

The next article is of the greatest importance, the means of preserving the mental and physical health of the operators and team members. The SWAT establishment must be careful to keep up to date in regard to case law and recent law enactments, as well as providing proper psychological service and physical training. The aspect of mental health can be greatly helped by personal recognition of the work the establishment does and by confirming the supervisors are using a system of positive rewards. As to constitutional and state law, as well as any case law, the operators and officers must know exactly what they may do as well as what they must do. They must be fully briefed and trained in the legal aspects of the performance of their duty. It must be insured that they are up to date on all search and seizure case law as well as use of force issues so as not to become civilly liable.

In the matter of psychological services, the SWAT establishment team leaders, supervisors, and commanders must be all alert to and constantly looking for signs of stress. They must also make sure that in every use of force the proper support personnel are in place to preserve the operators and other officer's mental health if needed. Finally, do not overly criticize or allow those with no knowledge of the SWAT establishment to unfairly judge their actions. All facets of the art of SWAT require constant cultivation to master; do not let anyone with no experience or extremely limited knowledge judge their actions and tactics.

It is the duty of the superintendents of the team to take care that operators who have become too old or are too injured to operate effectively and safely retire them from the team. Those operators who can no longer perform every duty must be reassigned for their own good as well as the good of the department, the SWAT establishment, and the public. The operators must pass regular physical fitness tests, solve complex problems, and prove they are able to react to varying situations effectively. For this reason the operators will exercise regularly and extremely hard, at all times and in all conditions they are likely to face in the jurisdiction served. This means that when they train under hazardous conditions, injuries to them are to be expected. Injured officers must be attended to immediately. Again, expect injuries. SWAT service is very much like any contact sport; there will be injuries even from training due to the hazardous nature of the duty.

Care to Provide Provisions

Depriving an offender of basic necessities will cause great havoc in the offender's morale and tends to be effective in inducing a surrender. This same equation can also work against the SWAT establishment. Time and opportunity may help to retrieve other misfortunes, but where basic provisions have not been carefully provided, the evil is without remedy. The main and principal point in SWAT is to secure and provision the establishment while denying these needs or trading them to the offender through negotiations for advantages. Exact calculation must be made before the commencement of a SWAT call-out as to the number of operators and officers and the expenses of the incident. This will allow for the timely delivery for all of the basic essentials as well as providing a means to deliver other supplies that may be needed to resolve the crisis in a timely fashion. These types of supplies must be in more than sufficient quantity and gathered into secure and efficient mobile dispensaries. They must be gathered and readied prior to the opening of a call-out.

If extraordinary amounts of supplies are needed, then requests for mutual aid via other agencies (such as the Red Cross, the Salvation Army, other community groups, as well as other policing agencies) must be sought out to fill the void. Purchase orders must be used and authorized in a timely fashion to procure all things requisite for the service.

On a SWAT call-out, the entire SWAT establishment should carry MREs with them as well as water. It must be realized that the heavy equipment carried, along with the physical exertions and stress induced in taking positions of advantage and performance of all of the duties will quickly dehydrate and cause a need for sustenance. The ICS team will be responsible for getting fluids and sustenance to the rest of the establishment. Additionally the operators will need periodic rests and the ability to stand down or be relieved. The greater the degree of variation in the weather, the more frequent these needs will need to be met.

During long standoffs, these needs become doubly necessary, even though the operators may not be doing anything for long periods of time. The mental stress of being on a call-out, as well as the physical stress of wearing and carrying the heavy loads quickly consume energy and fluids. On most call-outs the operators and officers will respond, resolve the call in just a few hours, and then take care of their own needs. On any extended call-out, the operators will bring supplies, but these will quickly be consumed. The purpose of the rations they carry is for emergency use. Your team must have a system in place to resupply these rations and meet your operators' needs. Even something as simple as not being able to properly hydrate just one operator will cause heat problems; this in turn will cause the team to lose other operators who have to help remove the now injured operator to a safe place. This will also result in hospital stays and expensive workman's compensation problems. It should be remembered that all members of the SWAT establishment will come to a SWAT event with varying amounts of sleep requirements.

In most agencies, at least some of the SWAT establishment will have work schedules that are at odds with the present call-out. Care must be taken to make sure that sleep deprivation does not alter team performance and that overly tired operators and team members are not deployed for extended periods. These precautions often become doubly necessary in offender barricade situations as they sometimes become protracted beyond expectation when the offenders resolve to suffer themselves all the inconveniences of want sooner than surrendering. If there are few hopes of the offender surrendering in a timely fashion, the establishment should immediately begin planning to convey the necessary sustenance to the SWAT establishment. They should also be taking all means to control the basic needs to the offenders.

During prolonged call-outs, the SWAT establishment must be constantly performing the basic maintenance their equipment and weapons require to maintain function. It may also be necessary to evacuate citizens from the vicinity of the call and provide a means for their basic needs to be met. Again the Red Cross, Salvation Army, and other local private groups can help meet these needs, as well as opening public buildings to house them until the crisis is resolved. It is imperative that displaced citizens do not become victims of the offender either through the offender's actions or the actions the SWAT establishment must take to apprehend the offender.

All members of the SWAT establishment will need to be supplied. Even the members of the command post and perimeter positions, as well as the negotiators will need to be given breaks. The needs of using a restroom for personal hygiene as well as supplying the entire establishment with sustenance are imperative. The establishment cannot leave the immediate crisis scene; if they use their emergency rations, then the rations are no longer available, and no one ever really knows how long a barricade situation will last, much less when it will occur. Take the precaution of having all of the sustenance preplanned and readily available. It is much less expensive to buy provisions prior to the need for use as well as being much easier to accomplish the supply acquisitions administratively.

Finally, be prepared and staffed to complete the criminal investigation and properly document and record all evidence after the situation has been resolved.

Prevent Arrogance and Keep the SWAT Establishment on Task

The SWAT establishment will need to prevent arrogance and encourage initiative to learn new techniques within the establishment. Every well-trained and disciplined team will grow to depend upon their tried-and-true tactics. However, the situations and community itself is constantly evolving. Technology and weaponry will continue to advance and will change the team's tactics as it becomes available. Arrogance prevents many teams from evolving. The team must be rewarded for looking forward, for adapting successfully to the communities threats as well as its needs. Reward inventiveness with the proper funding and allow for change in policy and procedure as necessary. Arrogance and an elitist attitude will cause failure and can eventually prove fatal on a call-out.

Make sure that the SWAT establishment is not overworked and is not taking on low-risk call-outs they should not be handing. The SWAT team should not serve every search warrant. Low-risk warrants and multiple uses blunt the keen edge of the team just as much as non use does. There must be protocol in place that defines which missions the SWAT establishment will accept to differentiate from those missions they must not accept. Only by continually reviewing this process will the SWAT establishment engage in the proper amount and types of work.

The operators should be disciplined just as the other officers in the department. If an employee can't perform basic police duties, they can't be in the SWAT establishment. Remember that at all times the SWAT establishment will be exposed to the community it serves. They should also participate in show-and-tell type events from time to time where they meet with the community and display some of their capabilities. This builds confidence in the community and helps to dispel the community from over reacting to their necessary actions later. This also keeps the SWAT establishment in touch with who they serve and puts them on notice that they must serve their community.

Approaching Offenders

The operators must arrive ready to quickly resolve any call they are dispatched to. Opportunities missed are opportunities lost. The operators must be dressed, equipped and deployed, the negotiators on scene with a plan of action and making contact with the offender, and the ICS must be in control and securing the outer perimeter as well as provisioning supply and equipment instantly upon arrival. The supervisors and commander cannot be too careful in making plans and keeping a reserve of operators ready to respond to a crisis point at a moment's notice. The SWAT commander must remain personally detached from the team supervisors and leaders' plans as they unfold and constantly be on the lookout for ways to improve or supplement them for success. Being detached from the plan decreases personal attachment to the operation and helps keep the operational focus in view. Let the team leaders and supervisors make the plans; the commander should then approve them or change them as circumstances require.

At the first opportunity, team supervisors and leaders should draw up and secure maps of the area to be searched or secured. They should designate all routes of ingress and egress and make note of any advantageous terrain, cover, or concealment that can be used to advantage against offenders. Operators should never rely upon the maps drawn up by others. Whenever possible, they should reconnoiter the objective themselves. The supervisors should go over their plans with each other before presenting them to the SWAT commander. All means of intelligence must be used to determine mapping, including city and county records, tenant recollections, and those of freed victims, hostages, witnesses, or otherwise known parties.

All manner of guard must be put on each point of egress to properly control them. The only way to ensure offenders will not escape is to have the perimeter supervisor emplace all inner-perimeter officers, and have an ICS supervisor personally emplace and frequently check on the status of the outer-perimeter positions. There is a tendency for officers and operators to lag in attention during long periods of inactivity on call-outs. Additionally there is a very real tendency for officers and even operators deployed with vehicles to sit inside the vehicles while on perimeter duty for comfort. This practice should be avoided if at all possible (excepting weather extremes) because the vehicle reduces the operators and officers' senses to detect offender activity as well as serving as a distraction or placing the officers or operators into relaxed a position and being unable to react quickly.

The entry and perimeter teams must make its approaches secretly and attempt to make entry into any enclosure with the utmost stealth. Keeping the general public from notifying the offenders by purpose or accident is of the utmost importance. Keeping the media from betraying plans and operational actions is also important due to the ability of an offender to monitor the news sources. A silent and undetected approach will surprise offenders and greatly enhance the odds of a successful resolution of the event. The use of pole cameras and throw phones can greatly enhance the use of stealth by providing current intelligence on the status and location of offenders. The scouts, snipers, and their assigned observers should continually monitor the inner perimeter and report back changes or dispositions of offenders and hostages as well as any other events worthy of being mentioned. It is generally safer for SWAT teams to operate at night and use night vision optics, or to use the cover of darkness for deployment and movements. Every opportunity must be taken to blind offenders. This would include deploying smoke and using background noise or creating noise to mask movements or distract an offender. Additionally offenders should not be allowed access to any outside communication into or out of the location they may be barricaded in. If the offender is known to be using a cell phone or walkie-talkies, the means of communication should be interrupted or otherwise compromised and listened to.

Once surrounded on all sides, the offender should be given the chance to surrender and repeatedly reminded that the option of surrender is open. The decisions to further limit or deprive an offender of space by use of irritants or operator movements should be carefully examined to determine if the benefits of such deployment outweigh the risk to the operators or public. Once communications have been established with offenders, the negotiators should take over these communications and begin laying the groundwork for surrender. Care must also be taken to deprive offenders with electronic contact with other offenders or persons outside of the perimeter. There are few if any circumstances in which an offender should be allowed to go mobile once surrounded. Take care to disable any vehicles the offender might have access to as soon as safely possible.

If the offenders should somehow attempt to go mobile, this should be stopped immediately with a use of force that is legal and meets the department's policy and procedure. Provisioning a ram vehicle, blocking vehicles in with heavy equipment, closing off roadways with barriers, using spike strips or otherwise disabling vehicles are standard courses of action. Offenders should only be allowed to get in vehicles that are under the direct control of the operators and that the operators have already prepared for an assault or to be otherwise stopped and easily disabled. All means should be used by negotiators to obtain a successful surrender, trading a vehicle for hostages insures the safety of at least some of the hostages, providing a vehicle can also set up the offender for an assault that will almost always prove fatal for the offender.

The supervisors are the masters of arms and control the SWAT team in the field. They must not initially post themselves at the front of any situation, yet they must remain in a position close enough to control all team movements and all operators' actions. Operators are expected to react to the offender until the SWAT establishment takes the initiative; once the operators have taken initiative, their actions should be closely guided and controlled by the supervisor. The position taken by a supervisor in any entry is usually no closer than third and no further back than seventh of an entry stick. The position on perimeter or in outdoor open field operations would normally be slightly to the rear and in the middle of most skirmish lines. When operating in standard military four-man type sections the supervisor should be in the middle position. The supervisor should be constantly considering their positioning so as to be able to react immediately as a back-up operator while still being able to control the other operators actions.

Danger of Movement

Being around any type of water in full SWAT gear is extremely hazardous. The weight of the heavy equipment and personal body armor makes it impossible for the operators to swim in most instances. Similarly, crossing open ground and entering long corridors, breaching doorways and approaching offenders are also the most hazardous times for the operators. All manners of conveyance should be used to make the arrival safe. Armored cars, covered approaches, the use of smoke for concealment and distraction by various means should be used whenever approaching an offender. The operators should use stealth whenever possible, and each operator should be putting themselves and readying themselves for positions of advantage to provide lethal force if necessary.

The use of smoke during the day and night has different applications. During the day, the use of smoke for concealment may also alert offenders to the position that is to be traversed. At night the use of black smoke can simply blind offenders. Smoke should nearly always be placed in front of the operators' movement. Distraction devices can also be deployed as a distraction to divert a suspect's attention away from a particular direction and/or to temporary disable their senses and give the operators time to enter, find, and engage an offender before the offender can react. Distraction devices emit blinding light along with an extremely high auditory explosion. These two combined effects will most likely temporarily paralyze an offender. The use of distraction devices in dark atmospheres can blind everyone; care must be taken so as not to blind operators. Remember the effects of the distraction device can be easily passed on to operators, and they must be properly shielded and alerted so as not to suffer themselves from its effects. Distraction devices can also be used to imply the use of force upon an offender and can be used as such to get their attention while also intimidating them.

Rules for Deploying the SWAT Establishment

The SWAT establishment should never train or conduct actions open to the eyes of the public or potential offenders if possible. Allowing offenders an opportunity to see the operators' tactics is folly. Show-and-tell sessions should be of equipment only and should never truly show all of the team's capabilities. The purpose of public viewing should be only to gain enough of the public's confidence as to cause acceptance of the use of the equipment. Any showing of equipment or operator movement on a call-out should be done only out of necessity or to intimidate offenders. Remember that the public and offenders have no need to know how the SWAT establishment operates; they need only know they operate very efficiently.

The use of television programs or encouraging news media to show SWAT tactics is folly. While it is true that most of the common tactics used are available in books (such as this book) or by reviewing news media coverage, the vast majority of the public are not offenders and never will be. Those who do become offenders generally do not plan well, so why allow them to know what they will need to react to? Whenever possible, keep details of operations confidential and maintain as much of a safe distance between the operators and the media as possible. Use public information officers extensively and stage any footage the media gets if possible.

The maintenance of the outer perimeter is critical as the officers therein must cover inwards and outwards. Thus a supervisor must regularly and frequently check on them to make sure they stay alert and do not expose themselves or the public to any hazards. Supervisors must be alert to the safe positioning of perimeter assignments. Remember that vehicles typically do not stop higher-caliber bullets, and most walls found in cities and towns do not stop bullets. Only large concrete barriers such as bridge abutments and larger building foundations will work as sure proof against the higher-caliber weapons.

One of the first things to be done when a SWAT call-out has been enacted is for the first arriving supervisor to confirm the inner perimeter is fully and safely manned. The supervisor should next establish a command post. Responding operators can then be used to contain the offenders, respond to risks, and minimize the risk to innocent bystanders and the sworn officers at the scene. The command post can then be expanded and special weapons and equipment handed out to the operators while the crisis negotiators begin their duties of intelligence gathering and determining how to contact the offender. Upon arrival of ICS, the outer perimeter should be finalized and inspected. Next the ICS supervisors should determine if further equipment or resources will be needed to resolve the crisis and begin the process of securing the equipment and bringing it to the scene as well as arranging for it to be dispersed.

The on-scene officers and any witnesses will provide the best intelligence to help for a plan on what will need to be done to resolve the call. The use of public records, public employees of all types, neighborhood residents, and anyone familiar with the area or event should also be identified and utilized to help form the plan to resolve the incident. Anyone entering or leaving the outer perimeter or crime scene should be thoroughly documented and interviewed for intelligence or knowledge for evidence later in court.

Motives for the Plan of Operations

As this abridgement is read, it should be remembered the ultimate goal of every SWAT call-out is a peaceful resolution, yet this is not always possible. Most SWAT call-outs are resolved in just a few hours, but some last for days and some have lasted for weeks. Every plan with the goal of a peaceful resolution should be considered as being potentially worthy, however, in most instances the tried-and-true methods and tactics practiced nationally will resolve most calls and should be used in most instances. Plans forwarded by persons of no or little knowledge should be looked at with a jaundiced eye. The SWAT establishment should be attempting to decline confrontation with the offenders if possible, yet when confrontation is inevitable they should resolve the incident quickly, efficiently, and when necessary, with a legal application of overwhelming force.

It is generally prudent to emplace the best and most-experienced command staff in positions of the SWAT establishment. This is not a place for someone to get on the job training. Command staff who are overconfident or never watch their operators train do not know their true capabilities. Command staff who do not know the state of the team's equipment and its location are also not to be considered for this position. The state of training and selection of operators is no more important than the tactics to be used or the equipment and supplies used to accomplish the resolution of a call-out.

The command staff must be weary of each and every call-out and the individual factors in whether or not to activate the SWAT establishment should be individually considered in each case. Each call-out must be examined to determine if a team truly needs to respond to it. Numerous questions should always be answered. Is this something that a non-SWAT specialty unit should be handling? Should the offender be arrested at this time or perhaps arrested more easily someplace else? Can the event be allowed to drag on, or does the risk of resolving the call endanger the community served? Is it possible that offenders who are heavily intoxicated or drugged may lose their will and eventually surrender or emplace themselves in a position of disadvantage? Should the team serve the warrant in the middle of the night to gain access when the suspects might be asleep and unable to arm themselves, or in the middle of the day when children would be gone to school from the area? Is it possible to make entry into the structure so fast and with so many team members as to flood the structure and instantaneously overwhelm offenders, or should just four to six team members make entry stealthily and then ambush the offender into surrender?

The team commander and team supervisors must be advised if possible in all manner of the attitudes, tendencies, and present state of mind of the offenders they are confronting. A complete psychological profile along with the criminal history and personal contacts and associates of the offenders should be prepared and considered when possible before a decision is rendered on how to resolve the call. Offenders with no or little knowledge are more easily dealt with than hardened criminals or offenders with prior policing or military experience. Violent desperate felons need not be pushed into corners if possible from which they perceive they have no alternative but to assault the operators, the public, or hostages.

The SWAT commander must know how far to depend upon the ability of the crisis negotiators and use the intelligence gathered by the scouts, negotiators, and command post. The SWAT establishment must determine the offender's confidence, mental state of mind, and intentions. Every attempt must be made to put offenders in positions of disadvantage tactically and psychologically. Any chance to further demoralize offenders and have them see the futility of their resistance must be taken. Only through seeing surrender as the only option can an offender be brought to arrest. Command staff must not lead or take close part in any SWAT action unless the most dire of circumstances dictate it. The command staff's role is to oversee the functioning of the SWAT establishment. Emplace the right supervisors in the SWAT establishment, give them direction, then let them do their duties.

Do not let inexperienced operators and negotiators be placed in roles where they are not supported by highly trained counterparts. Operators unused to intense decision making should not be put in lead roles. Negotiators who have not had extensive training should not be placed in the lead position. It is a good practice to have all of the SWAT establishment assembled twice a year and have them train together. Otherwise the operators and the rest of the SWAT establishment should train separately in every part of their drill by their supervisors who are masters at that skill set. The SWAT supervisors and team leaders should also be the weapons masters and lead trainers. In this way the supervisors will know their skill and strengths, be able to see the operators perform their evolutions with proper regularity, and to see that they are sufficiently attentive to voice commands, hand signals, and able to obey orders. If any operator or officer is deficient in any of these particulars, they must be instructed and exercised until perfect. Practice does not make perfect. Perfect practice makes perfect.

Although the SWAT establishment may be thoroughly disciplined and complete in all field exercises, in the proper use of handguns and long arms, in the evolutions of dynamic and slow and methodical searches, it is not advisable to lead them rashly into any call for use. The call-out criteria previously mentioned must be adhered to, and the operators must take on roles of certain responsibility only after multiple calls for service have been favorably resolved. A vigilant and prudent supervisor will carefully weigh in his mind the state of each operator and the individual factors of the current call-out before making team assignments. If the SWAT supervisor feels the call is resolvable safely and that younger, less-experienced officers can be put into more demanding roles, then a newer recruit can take on additional responsibility if a more-experienced operator is positioned to immediately assist them. If the odds or situation dictate extreme danger and decision making along with requiring use of all skill sets of an experienced officer, then newer operators should be given lesser roles or roles with experienced operators in position to immediately assist them. In every case, the supervisor should be using stratagem and attempting to place the team members in positions of advantage, using stealth over force whenever possible, encouraging surrender, and placing the offender in position where surrender is the only option.

What to Instill Upon the SWAT Establishment

It has been said that is better to be lucky than good, but make sure that you are damn good when your luck runs out. Do not rely on luck when there are the time and resources available to make your SWAT establishment good. All skill sets and disciplines are to be brought to perfection through continual practice. The SWAT establishment is still responsible for protecting the rights of all they come in contact with as well as enforcing the law. They are the means by which all civil liberties are preserved, dignity is perpetuated, and the public's trust is maintained. This truth must not be sacrificed. Any department that fails to respect the rights of all involved, victim, witness, suspect, and innocent bystander confers no service at all. In short it is indispensably necessary for those engaged in SWAT actions to not only instruct their officers in the means of preserving their own lives but also how to use force judiciously and prudently. They must protect all constitutional rights and especially the rights of due process. By keeping to this maxim, the SWAT establishment will never fail in its mission.

The department's superintendent must know that their power and dignity are so great and their fidelity and bravery are the fortunes of their department, the public they serve and even the lives of their officers and families. The reputation of the department is entrusted to them, they should not allow any unit or part of their department to take on missions they are not equipped, trained, or intended for. If the SWAT establishment cannot handle the call, then it should be passed on to an agency that can either assist or handle it themselves. Similarly, no SWAT establishment units within the department should not take on missions of greater risk or ability than they hold and can handle themselves.

The SWAT commanders should know the abilities of each and every operator, negotiator, and member of the ICS team. The commander should assume respectable authority and maintain it through professionalism and competence on their own part. The commander must be willing to punish any member of the SWAT establishment appropriately with the intent of keeping the integrity of the SWAT establishment beyond rebuke. The commander must look for every opportunity to instill positive character in their subordinates and endeavor to set the proper example of the SWAT establishment and themselves to the public.

Having once firmly established these regulations, the commander and supervisors must watch for moments of opportunity when offenders are dispersed, inattentive, placing themselves unwittingly into positions of disadvantage, or having separated themselves from hostages, or from their weapons, and order the team to take them into custody or otherwise resolve the event. The veteran operators and negotiators will recognize the moments and report them back; if acted upon, this will instill and encourage initiative in them. It will also lead to multiple opportunities to resolve the event. The SWAT establishment will be maneuvering the offender into ambush or otherwise be using surprise and stealth to surprise the offender. They will use concealment and cover to move upon offenders and deprive them of space, while simultaneously attempting to compromise the offenders' ability to use cover and concealment.

Typically, most SWAT assaults will come when offenders are asleep or otherwise most vulnerable, tired, and exhausted emotionally and physically, and/or distracted. This should be the norm for most SWAT plans of action, if cover, concealment, stealth, and surprise are not be used to supplement and minimize the use of force, then the plan of action should be reconsidered. The SWAT establishment's goal is to prevent the loss of life whenever possible and to minimize the use of force. To do otherwise shocks the conscience of the public and turns them against the agency and policing as a whole.

It must be remembered that utilizing overwhelming force does not mean injuring offenders. Presentation of and emplacing the overwhelming force into position frequently induces offenders into surrender. When this fails or the offenders' actions demand it, overwhelming force must be used to immediately stop the offender from injuring others. The SWAT establishment should be doing everything reasonable to take offenders into custody without using force. They should also be thoroughly prepared and ready to use force to stop an offender and take them into custody. Finally, the SWAT commander and supervisors' plan to resolve the incident must use force to resolve the incident only if surrender and compliance, or the imminent threat of bodily harm to others is being done or will most likely be completed by the offender.

If an offender attempts to flee or attack operators,

the offender should be immediately dealt with. There should be snipers and perimeter operators in place to respond immediately with the appropriate force. The commander should have operators emplaced at all avenues of escape and in all positions that might yield sight into the event. Additionally all offenders should be engaged at as much distance as is practical in order to lessen the offenders' chances in marksmanship and increase the operators' odds of survival and successful application of force. The SWAT operators will be highly trained in marksmanship, and they should be placed with the purpose of using it to the greatest advantage. Prudent commanders and supervisors will ensure that spike strips, door stops, reduction or introduction of light, and other means are used to restrict, hinder, or funnel the movement of offenders. Prudent commanders and supervisors will also rely heavily on the crisis negotiators to tire and demoralize offenders while recognizing the time available to do so may be dictated by the circumstances.

One maxim that must be remembered throughout this work is that no one should ever despair of effecting what has already been performed. As an example, if the operators have prepared for operating in a gas environment, then there should be no hesitation of using gas when appropriate. There should be no great worry about the marksmanship of the operators if they have performed their qualifications at the 100 percent level. There is no great worry about small arms fire taking the life of operators if they wear body armor. There is also no need to worry about the ability of the operators to use cover and concealment if they practice it regularly. Crisis negotiators will do their assignments proficiently if they have been trained to proficiency.

ICS personnel will maintain the outer perimeter and secure logistics if they have been properly trained.

Again, the practice of deceit and stratagem is paramount to ensuring an appropriate use of force or a surrender. Nearly all offenders when cornered will barricade themselves and attempt to take positions of advantage. The use of stratagem, slow and methodical searching, gas, robots, technology, and canines allow the operators to search safely and stealthily with a minimum risk of danger to themselves, the public, and the offenders.

The SWAT establishment should not be afraid to learn or fail to learn from others that with which they have no experience. The National Tactical Officers Association and Crisis Negotiators Association regularly extend affordable training opportunities that include guest speakers and instructors from around the United States and the world. These experts provide excellent current realistic training and impart advice on their own mistakes as well as their successes. The operators, officers, supervisors, and commander must be sent to this training. They must also be willing to accept and adapt advancement in the art of SWAT from other agencies and departments.

The SWAT establishment must collect data on the successes and failures in its critical events. The SWAT establishment must be further willing to honestly debrief each and every incident. The establishment and the agency itself must then be willing to share informally within and without the agency to others the proper lessons learned from these incidents. After the proper time has passed and the agency is no longer liable to civil liability then the hard lessons learned should be shared formally with the rest of the SWAT community. In short we can learn not to make the mistakes of others, and we can learn how to more easily arrest offenders from new novel techniques employed by others than those within our own agency.

Preparations for General SWAT Actions

After having explained the branches of SWAT establishment, the order of policing naturally leads us to the SWAT action. The SWAT action itself is a conjuncture and can be fatal to municipalities and agencies as well as offenders, officers, and the public. The decisions made during a pitched action will determine either the fullness of success or failure. This eventuality above all others requires the exertion of all the abilities of the SWAT commander, as their conduct on such an occasion gains great glory, or endangers the entire SWAT establishment, the community, and the offender. It is at the moment of the SWAT call-out itself in which the commanders' talents and experience show themselves to their fullest extent.

The SWAT establishment should formally brief all involved of the nature of the event and the reasons the offender is going to be arrested, so that any subsequent use of force will be known to all as justified and falling with in policy and procedure, and more importantly the law. Even the simplest briefing allows the operators to advance with confidence and know that the use of force option they select will comply with the needs to resolve the event. Not having the benefit of a briefing on the actions taken and being taken by the offender encourages tentativeness upon the part of those trying to resolve it. Being tentative causes mistakes and passes initiative to offenders. Mistakes in SWAT actions can be fatal and are hard to retrieve without loss.

Proper measure must be taken during static events to keep the entire SWAT establishment current on the unfolding events of the action at they occur. Changes in the offender's position may compromise an operator or present a moment of opportunity. Changes in the mental state of an offender may signal the need for the operators to be prepared to instantly resolve the call; likewise they may also give them time to stand down and preserve their energy. A SWAT operator must not be continually used on a static event. No person can maintain their position or post for more than sixteen hours at a time. The team commander must be willing to request that the incident commander call in other resources in order to properly rest the SWAT establishment.

Remember that an operator's strength can be dissipated just in movement or other simple actions. The carrying of the basic SWAT load of firearms, protective equipment, and other various general issued equipment will tire an operator over time. Additionally, high stress levels can lower their ability to remain alert, perceive danger, respond in a timely fashion or take initiative at a proper time. If possible the operators should be rotated into positions where they can stand down, relax, and regain their wits and strength. No offender, wearied by constant harassment and harangue of the entire SWAT establish can stand up to the actions of a fully alert, well-rested, and properly disciplined SWAT team.

Sentiments of the SWAT Establishment

It is necessary to regularly train and remind the SWAT establishment of the departments policies and procedures as well as the law. They must be continually reminded the true art of SWAT is being able to use the least amount of force available to make an arrest. This is because they will continually train to use force and must not take this part of their training to mean that they will always use force. The SWAT establishment must also remember that they are police officers, will testify later, and will be responsible for securing evidence, witnesses, and crime scenes. Finally, they must remember they serve the public and the offenders, and their duty will be scrutinized to such a high degree that they must continually observe every protocol in order to keep the public's confidence. The SWAT establishment should employ every means to remain capable of never showing rage, hatred, or indignation toward offenders, especially when in public. Training should reflect this, and the commanders' own actions should also reflect this, especially to their direct subordinates.

It is natural for all officers to be affected with the sensations of fear and to have subsequent emotional responses after a SWAT action. Strict training and discipline will minimize the effects of the emotions during the SWAT event. However, the resolution of the event and the exposure to high stress and the results of violence can take an emotional toll. When the SWAT establishment successfully resolves a call, the superintendent of the agency should show them approval with a public recognition of their successful service. The team itself should be recognized, and if individual bravery and initiative was exhibited, it should be recognized too. The SWAT establishment should have access to the other support services in the department and encouraged to use them. The use of critical incident debriefs will help provide an excellent means for individual officers and operators to speak and find solace for their actions. Psychological counseling through paid psychologists and the use of other officer based groups to help the team vent after particularly stressful events should also be encouraged. The use of psychological and readily available support services will keep good, trained officers available for service and extend their SWAT careers.

Increasing the Odds of Success

Good SWAT commanders and supervisors are acutely aware that success will depend much on the nature of how much advantage the team can take from cover and concealment, as well as recognizing opportunities to create surprise and compel a surrender. High ground (if available) should always be secured and used to provide snipers and their observers with good lanes of vision as well as a means of gathering intelligence. Cover is any item that is proof against all small arms fire. Concealment is not cover; concealment is anything that keeps the offender from seeing the positions of the operators. Concealment can also be in the form of light or the lack of light, anything that may obstruct the offender's vision can be used for concealment.

Order of Deployment

When emplacing the SWAT establishment to contain an event, three things are to be considered: the weather, critical terrain, and the amount of time there is available to resolve an event. The weather will affect how long the establishment of operators and outer perimeter officers can be safely deployed. Intense cold or heat can deplete an officer physically in minutes. The SWAT commanders must be concerned with keeping all of the establishment away from the dangers of the elements. Additionally the team should be taking advantage of the elements. Use of darkness for concealment, or using the sun or artificial light to help blind the enemy is considered standard operational procedure and should be a consideration to aid the team. Leaving an offender exposed to the elements can also help wear down the offender and tire him out, so much so as to sometimes bringing on a surrender of its own accord.

Proper Distances and Intervals

Having explained the general dispositions of deployments, it is also important to consider distances and dimensions. The operators should always be placed in positions of cover or concealment. The placement of officers in the inner perimeter will be dictated by cover and concealment available. The placement of the outer perimeter and command post will also be dependant upon concealment and cover available. Outer perimeter positions must be free from direct line of sight of the offender. The officers deployed in the outer perimeter must be able to move around freely without fear from harm from the offender so they can keep the public from being harmed. Finally, most perimeter positions will not always be on the drawn lines in the map room. There must be enough officers deployed to the outer perimeter to allow the officers to see each other and everything in between them. If they can't see each other and cover every avenue into and out of the inner perimeter, the perimeter is either to big or needs more officers. Handguns are generally employed in the five- to ten-yard range. Long arms are generally deployed out to two hundred yards effectively.

Disposition of the Command Post

The command post should be set up in an area that allows easy access to thoroughfares, water and restroom facilities. If electricity can also be secured this enhances the command post's position as it will allow for not having to use emergency generators. Medical trauma units should be placed here or with the outer perimeter personnel as the situation deems fit, if their use is immediately anticipated they can be moved inside and positioned inside the inner perimeter. All outer perimeter and command post personnel should have easy access to vehicles to allow for a mobile response should the situation somehow develop into a dynamic moving situation.

If the SWAT establishment is part of a small department, there are limited resources, or the situation demands deployment through lack of area then it is acceptable to deploy all resources at the main command post. However, if the situation is very large or is requiring a large deployment of resources then the ICS team should separate all of the different resources into groups in nearby assigned areas. This will allow each unit to avoid the friction of the other units and allow them to deploy and be ready to meet an evolving situation.

Reserves

The ability to deploy a reserve away from the inner perimeter for a quick response may not be one that the SWAT establishment or the situation allows to occur. In any instance, the SWAT operators must be ready and able to perform any SWAT duty, including perimeter, entry, or sniper duties. If the SWAT commander has the ability to deploy a reserve, the commander should do so. Having even a small group of one or two officers or operators ready to quickly respond to any emergency or forced action by the offender will allow filling any vacancy or act as a reinforcement to a sudden change in circumstances.

The reserve's duty is to supplement and if necessary take direct action to resolve a call at a critical moment. These operators or officers should be prepared to act as an arrest team if necessary, deploy diversion, or use any level of force. They should have the means to have immediate access to the critical area, thus if a perimeter is small, they may be deployed near the offender, while a large perimeter may require them to have access to ground vehicles or air assets for transportation. The reserve should have talented operators who are properly armed and protected with body armor so they can meet any developing situation.

The Post of the SWAT Commander

The position for the SWAT commander should be near the ICS command post. The SWAT commander should not make plans with the SWAT supervisors and team leaders inside the incident command post unless lack of space indicates placement there. SWAT commanders must be free from the distractions of the operation of the command post to allow them to stay focused on resolving the crisis and make instantaneous decisions to help resolve it. The SWAT operators and officers deserve confidentiality in their actions and to be free from the media broadcasting their every move and anticipated move. Positioning the SWAT command post near the ICS command post will allow the SWAT commander to access all resources needed to resolve the call-out as well as confer with experts and departmental command staff upon the progress of the mission.

Once a SWAT call-out has been initiated, the SWAT team operators should be inserted into positions in the inner perimeter as soon as possible. If the situation was one that was not pre-planned, then officers from the inner perimeter should be replaced with SWAT officers when reasonably possible. These officers may then be debriefed and redeployed if necessary to the outer perimeter. The use of background noise, artificially created noise, and use of light or other distractions should be considered to assist in the safe deployment and extraction of officers and operators.

As soon as possible, the SWAT supervisors should perform a reconnaissance and intelligence review of the situation. They should immediately plan a response to resolve the incident. This plan should contain contingency plans with a force response appropriate for the current situation. The SWAT supervisors, ICS, and perimeter supervisors should review and confirm the positioning of all inner and outer perimeter personnel to confirm they are in locations of safety that provide concealment and/or cover from offenders. Knowing the exact positions of all operators and officers will allow for the proper economy of force later to resolve the situation.

As soon as operators are in a safe position they may consider and implement the use of noise distractions to purposefully intimidate, distract, or confuse offenders. Specialized SWAT equipment such as helicopters and armored cars and deployment of diversionary devices and irritants can also be used for these purposes. This show of force can serve as a distraction as well as give the operators the initiative and possibly demoralize offenders. Effective use of equipment in this manner can also consume the offender's ability to reason out a response or create an effective counter to the arrest efforts. Creating confusion in the offender creates the ability to get inside the offender's response curve and will give your SWAT establishment the greatest opportunity to safely resolve the call-out.

Maneuvers in Action

Opportunities missed are generally opportunities lost when dealing with offenders. SWAT supervisors and crisis negotiators should be looking for every opportunity to deprive offenders of rest or time to reason out effective counters to the efforts to arrest them. Additionally the offender must never be allowed to truly believe they can control any events accept a peaceful surrender once the SWAT team has arrived. Any chance or opportunity to separate offenders from potential victims must be exercised as soon as reasonably and safely possible. Every opportunity to ambush the offender into surrender must be explored and recognized. At some point the supervisors must be willing to accept some risk in order to resolve the situation; however, the supervisors must be doing everything possible to make any engagement with an offender unfair; they must stack the odds of success greatly in the operators' favor. This means the supervisors must be constantly considering stratagems of deception and subtlety to resolve the situation. The action may or may not require a use of force, but the supervisors must consider all options to resolve the call.

The supervisors' first duty is to ensure the safety of the uninvolved bystanders and the general public who have not been victimized. This means that their planning should be directed toward containing the offenders and situation into a confined area that allows the SWAT operator's team to keep the offender in and the bystanders and public out.

The SWAT establishment's next duty is to the safety of the responding personnel, followed by ensuring the safety of any hostages and/or victims. There should be carefully thought-out plans and drills used by all officers as they will in fact be placing themselves between offenders and the public's safety. However the supervisors will need to be devising stratagems that maximize the chances of success and officer survival while accomplishing that goal. The goal includes getting all of the officers home safely, as well as ensuring public safety. If the officers are injured or killed in the line of duty today, they can't protect the public tomorrow. At some point, a degree of danger, even great danger for your officers, will have to be accepted to resolve a call, but this does not mean that hostages and victims lives are traded for those of officers.

Generally there are two types of SWAT actions, static actions, and dynamic actions. In static actions, the SWAT establishment has total control of the initiative and is controlling the offender's actions and movements. In a dynamic action, the offender is attempting to take initiative, and the SWAT establishment's actions are not necessarily dictating the offender's actions. Whenever there is time, Whenever there isn't time to make a plan, make a plan. Failing to plan is planning to fail. Constantly planning and planning for contingencies should be practiced throughout the SWAT establishment. "What if . . . ?" should be constantly being reviewed by each member of the SWAT establishment. In this manner, the result of all officers' actions are known and yields a successful resolution to the crisis.

The SWAT team must be ready to be deployed to static actions, or enter into a dynamic situation at any time. Generally SWAT calls are initiated after offenders have taken refuge or been identified as being inside of a confined space or area. SWAT call-outs in static situations are generally resolved by opening lines of communication with the offenders through the use of skilled crisis negotiators. Static situations with barricaded offenders and no victims can be pushed to a conclusion by the use of distraction devices and irritants to help the offender appreciate the helplessness of their situation. Use of irritants also helps "blind" the offender and increases officer safety as the operators should be well versed in the delivery of irritants and operations within irritant environments.

Deployment of less lethal options to resolve a call-out during these types of operations are generally well planned and implemented at the direction of the SWAT commander. This is what is meant by the imposing of the SWAT supervisor's will upon the offender. The offender will either surrender to arrest or be forced into arrest. In the static situation, the SWAT establishment is controlling the actions of the offender. The static situation gives the greatest chance of a peaceful resolution of the SWAT call.

In a dynamic situation the use of less lethal options to resolve the crisis can arise and be implemented, but the use of force is at the discretion of the operators in the field and confronting the situation. During dynamic situations the potential for the use of lethal force is always higher due to the actions of the offender not being rigidly controlled. The operators and officers will be responding with force that matches the department's use of force policy and the law. It is important to remember that less lethal options are frequently not viable to resolve incidents with violent offenders who are embarking upon or taking potentially violent actions.

SWAT call-outs are generally fluid and will change between static and dynamic. This is almost always due to the actions of the offender, however skilled crisis negotiators working with well thought-out SWAT response plans, along with the proper economy of force will successfully resolve the call-out and minimize the amount of dynamic response by the offender. Above all, the SWAT supervisors and operators must not be afraid to properly apply force to resolve the call-out. This means command staff must be willing to allow them to apply their trade properly without interference. The SWAT establishment will be creating disorder and confusion to break the offender's will to resist. This will also be done to increase their advantage should a use of force become necessary. If a superintendent insures the SWAT establishment is well trained and disciplined then they will have done everything they could before using force. In short, the offender brings the result upon themselves and actually dictates if any force will be used on them.

Various Formations for Action

There are typical deployments for SWAT responses; however, the changing and differing circumstances of the every situation will dictate the actual deployment areas of operators and resources. Most of the entry team will likely be equipped with distraction devices, handguns, sub guns, and sometimes long arms, as well carrying a less lethal option of some type, usually baton or handheld irritant and necessary breaching equipment. The perimeter team will most likely have long arms and ranged less lethal weapons. Specialized personnel such as explosive breachers, grenadiers, and canine handlers will frequently be deployed in perimeter position or with the entry teams as the situation dictates.

The outer perimeter officers will most likely be uniformed personnel, although sometimes plain clothes detectives may be forced by necessity to assume assignments. Standard patrol-issued weaponry and protective equipment should be deployed with these officers as well as making sure they are properly identifiable and easily recognizable as police officers by everyone. Only uniformed personnel should be allowed inside the command post. This will reduce confusion and the chances of civilian bystanders or the media from learning of operational plans to resolve the call-out. It also provides instant recognition for command personnel for the dispatching of officers for arising needs.

Rarely, military personnel such as National Guardsmen are required to resolve extraordinary circumstances such as large on-going riots. In general, the use of military personnel to resolve dynamic situations or take positions in perimeters is discouraged. Military personnel are generally not trained to deal with civilian emergencies and problems, nor are they typically allowed by law to do so. Military personnel and involvement should be limited as much as possible to use of equipment assets and guidance in the use of those assets.

Fire personnel and emergency response (medical) personnel will inevitably be required at some point on almost every SWAT call-out. Trained and well briefed SWAT operators must be deployed for their protection and to guide them into positions to safely while they perform their duties. Some teams use and train with team paramedics. This allows for deployment of team members skilled in medicine to inner perimeter and even to entry positions. This does not supplant the need to be ready to work with other EMS or fire personnel. At some point, injured suspects and victims will be evacuated or transported from the arrest scene. Operators and all officers involved must be trained to safely interact with them.

Occasionally civilian equipment and assets may

be needed to resolve a crisis. Whenever possible, other government agencies and their assets should be explored before using civilian resources. The use of these types of assets must be strictly controlled and completely justified before being used. As much as possible, these assets should be under the direct control at all times of the SWAT establishment to allow for complete accountability and control during their use.

There is one rule that should be observed at all times and that must be reiterated. SWAT teams resolve SWAT call-outs. Narcotics units, gang units, and other specialized team units with limited training should never be used in lieu of a SWAT team. An established call-out protocol must be part of the agency's policy and procedure. Successfully resolving a SWAT action is gained by a disciplined and well-trained team. All potential SWAT actions must be considered carefully. The SWAT establishment should not be allowed to take on missions they are not trained for, and specialty units within a department should be held to the same standards. To not do so is folly.

Dictate the Offender's Actions

The use of crisis negotiators is imperative, and interaction between the crisis negotiators and the SWAT supervisors should be part and parcel on any call-out. Whenever possible offenders should be induced to surrender peacefully, without any use of force. This is most often not the case. Do not intentionally set or allow friction between offenders and the SWAT team. Utilize negotiators whenever possible and begin implanting the seeds of surrender to the offender. Allow the offender to believe and build the belief in them that a peaceful surrender is the best solution for them. Stay away from ultimatums during static situations if possible. Do not give an offender time deadlines, and seek out their demands. By seeking their demands, their resolve is weakened, and they may be placed into a position of disadvantage.

The SWAT operators should be maneuvering and exploring the means to use all force options to arrest the offender. The crisis negotiators should be using every means and stratagem to induce and weaken the offenders resolve into surrender. Failing that, the negotiators should then be utilized to place the offender into a position of compromise so that a force option can be used to resolve the situation.

Some Actions Should Not Be Undertaken

Not every request for a SWAT call-out is honored simply because of the request. Crisis situations should be immediately reviewed by command staff including the SWAT commander, and informed decisions can then be made as to whether to respond or not. Not every armed and barricaded situation needs to be resolved by a SWAT call-out. As an example, if a person is suicidal and barricaded in their own home, yet they have made no threats to anyone else and are not a threat to anyone else, then a full SWAT call-out generally should not be used to resolve the situation. The SWAT commander must be willing to decline as well as accept SWAT deployments. In order to safely do so and remain free from civil liability, the use SWAT of call-out protocols will be referred to. Only the accumulation and careful consideration of certain factors mandate deployment of a full SWAT call-out. Some situations will easily meet criteria and be "no brainer" decisions. Others will require exploring whether or not the quantifiable standards have been met. In any case, the justification for the call-out is easily defended later with the public and in court.

A rash deployment of the SWAT establishment will expose it to danger. Responding to improper call-outs puts the team in a quandary; they have been trained to respond to high risk situations and utilize all of their training and skills, yet they are being deployed when they are not truly needed. The increase in the use of resources and equipment subjects it to wear and equipment fatigue. Being deployed in non call-out situations leads to inattentiveness or places the operators in position of being used again and being tired from prior use. The risk of wearing and carrying all of the equipment, as well as unnecessarily exposing them and the agency to public scrutiny, are unnecessary and harmful. Remember the SWAT establishment is uniformed and equipped to intimidate with its presence as well as its actions. If there is no need to deploy them and cause fear in the public they serve then don't do so.

The best view of the SWAT actions for your agency are those in which the media and the public gets just a fleeting view or image of the SWAT establishment. They see only that the event was resolved and a dangerous situation was resolved. If possible, do not allow the media or public to see a necessary use of force or intimidation. Thus the public and media are not intimidated and fearful of the team.

Greatest Danger to the Operators

In the moments immediately prior to or during the time that the offender is being arrested (in the immediate vicinity and most likely physical contact), the SWAT Team is at its greatest risk. First, the offender will most likely know exactly where at least one of the arresting team members is. Second, the officers actually making the hands-on arrest must partially compromise themselves to make the arrest by touching the suspect and physically taking them into custody. Third, this is the time of most mental strain and stress. The adrenaline and pulse of everyone involved is usually raised at this time, and this is when a lot of physical confrontations occur. All operators should be well trained in the use of cover while making arrests as well as the surrender ritual. This will ensure that everyone involved or within listening distance knows exactly what is being done. The surrender ritual allows everyone involved to know exactly what the responsibilities of the offender are to ensure a safe arrest. Well-trained officers will say the same thing every time and use language the public will accept and not be offended by. This set use of language, if changed, signals everyone present that a use of force is either imminent or occurring.

The Surrender Ritual

The surrender ritual consists of an officer verbally identifying offenders and making sure the offenders and other officers know that suspect(s) are being confronted. The officer gives out verbal orders in a loud and demanding voice. The officer identifies any offenders as "suspect,"and order the offenders to submit by placing their hands in the air, to lie down on the ground on their stomachs, place their hands palms up far out to their sides, and finally to look away from the arresting officer. Further instructions are given as necessary until the offenders are taken into custody. Verbal warnings are also given to the offenders as to repercussions that will occur for not following these orders. Typically the orders given are as follows:

"Suspect put your hands up, suspect get down on your knees, suspect get down on your stomach. suspect, hands out to your sides palms up, suspect look the other way, don't move; don't make me shoot you!"

Immediately after an arrest, the danger to the operators escalates at a time in which the officers and operators want to relax. Good team supervisors and team leaders will not let this occur. Positions of cover and concealment must be maintained, as well as positions in the outer perimeter. Frequently second unknown suspects may lurk in the area, and measures must be taken to make sure that they are safely accounted for. Additionally, offenders sometimes take this time as an opportunity to assault officers by biting, kicking, spitting, or attempting to use body weight to assault them, or even attempt a belated escape. Inattention at this point can lead to an injury or unplanned use of force. Finally the outer perimeter must be maintained to preserve the crime scene for the follow up investigation.

Vehicles

Offenders cannot be allowed to go mobile by using vehicles or otherwise leaving the containment of the inner perimeter. The deployment of spike strips, police cruisers, or other vehicles and barriers, as well as keeping a potential ram car will render escape by vehicle useless.

Vehicle Assaults

In some instances, the SWAT team will have to resolve a crisis with an offender inside of a vehicle. In these types of instances, if negotiations are not productive (or feasible), the SWAT team must assault the vehicle. Whether the assault is of a smaller vehicle such as a sedan or of larger one like a bus, certain tenets are universal. Use of distraction to gain the operators the advantage of reaction time, use of dedicated breachers to gain vehicle access, control of practiced lethal force angles by designated operators, as well as designated vehicle entry and potential rescue operators, with all roles being pre-designated, practiced, and controlled via a team supervisor and team discipline. Finally consider the use of precision-scoped weapons to initiate the arrest and provide instant safety for the SWAT operators and any possible victims.

Open Field Searching

All teams, whether rural- or urban-based, must be practiced in open field searching, using cover, concealment, and maneuver by elements, orchestrated by team leaders and supervisors to safely cover expanses and search effectively and efficiently. Operators who move with alacrity and dexterity keep the risk of danger to a minimum, as well as ensuring the purpose of locating hidden offenders and taking them into custody. Well-conducted searches reduce the use of force by locating and compromising offenders prior to them reacting to the presence of team members. The use of sniper/spotter teams who are searching from a distance and able to impose lethal force or otherwise identify threats to the search teams is also recommended whenever possible.

Remote Controlled Vehicles

Remote vehicles (robots) are extremely valuable and should be used if practicable for primary entry prior to entry by live officers if possible. The use of a remotely piloted vehicle increases the odds of a safe entry into a location by reducing the exposure of the operators to offenders hiding in ambush. It must be remembered that any entry with a remote vehicle will ultimately have to be followed by entry by live operators. This entry must still be conducted by the operators with the same safety protocols as if the primary robotic entry was never made. Remote vehicles and robotic-based camera systems have limited fields of vision and cannot be considered absolutely reliable. They also cannot view every possible location that an offender may be lurking in. Just as with any piece of technology the team possess, they should be ready to complete the assignment without it. Batteries fail, equipment breaks, malfunctions, or doesn't function to its fullest capacity. The team must be prepared and practiced both with and without advanced equipment.

Rehearsals

Whenever a SWAT action is being resolved, the method of resolving the call should be practiced or rehearsed if possible. This rehearsal may consist of merely having the operators repeat their roles back to a supervisor to as much as rehearsing or practicing on replica vehicles, mock ups, or buildings, set as much as possible to resemble the incident scene.

It is of the greatest benefit to have SWAT operators rehearse entering and exiting vehicles and to have them practice putting on all of their equipment under duress. This benefit extends to crisis negotiators setting up all of their equipment and ICS personnel setting up command posts and practicing responding to outer perimeter positions.

Calling for Aid

While the SWAT establishment should be ready to attempt to resolve any situation, the SWAT commander must also be willing to accept and admit when the crisis is outside the ability of the SWAT establishment to safely resolve alone. At times the use of crisis negotiations takes time to deliver full positive results. The SWAT establishment's ability to continue efficiently and safely will depend upon the weather and elements it is exposed to, as well as the sheer amount of time itself. No team should be deployed without a break for longer than sixteen hours except under any of the most dire circumstances. The SWAT commander must be ready to request that the incident commander call in teams from adjacent jurisdictions so that fresh, alert officers are ready and in the best position to resolve the call.

In some instances, perimeters and areas to be searched are extremely large, or the call-out will require skills your team does not possess or has not practiced to a high enough level of expertise. Call in a team able to aid or resolve these types of crisis. The SWAT team should have practiced and trained the specific skill sets needed to resolve every call-out they accept. Discipline isn't just being willing to respond and attempting to act as ordered appropriately; discipline is obtained through practiced response.

No tactical plan should include accepted injuries to operators. The reality of the hazards of a SWAT action and the harshness of SWAT training is that SWAT operators run a substantial risk of be injured doing the jobs. It is a risky and hazardous undertaking. Plan for these injuries and deal with them by providing paramedics within moments of the crisis points.

Crisis Negotiations

Most call-outs are resolved through opening lines of communications. Practiced, well-trained crisis negotiators are the best option for resolving most SWAT actions safely. Crisis negotiators speak to the offender and address their concerns while incrementally imposing the will of the team commander upon the offender. Crisis negotiators frequently use reason and compel the offender to a peaceful resolution of the crisis. They guide the offender to the ultimate reality that surrender is the only recourse. The crisis negotiators are also to be used to constantly tie up the offender, distract them, tire them, demoralize, and ultimately guide them to a safe arrest. Even in instances where offenders will not surrender the utility of a crisis negotiator can be immeasurable. Crisis negotiators can be used to gather intelligence to take advantage against the offender as well as potentially placing the offender in a compromised position.

Standing Down

Immediately after any SWAT action, the entire establishment should meet and debrief. This will be used to address all concerns of the action and allow the SWAT establishment to provide additional information to investigators. Equipment performance and most importantly personnel performance will be critically evaluated while the incident is still fresh in mind. Additionally this is the time when supplies and equipment are made ready for the next action. All equipment should be put back into working order for the next call-out. Most importantly the incident is shared within the SWAT establishment so as to pass experiences on to other officers and to provide a basis for the need to alter or change tactics in the future.

General Maxims

All that is beneficial to the SWAT establishment should be detrimental to the offender, and the team should be taking steps to deprive the offender of any comfort or advantage. Cutting off water, food, electricity, and lines of communication to the outside world and potential co-conspirators is considered extremely important. Giving these types of comfort back through crisis negotiations conversely does not actually give comfort to the offender. Giving these items and necessities back actually imposes the will of the SWAT commander on the offender and places the offender's basic needs at the will of the commander. These needs are always at a price, even if the price is just talking to the negotiator.

Be cautious of turning off basic utilities to deny them to offenders. Turning off all of the electricity for the block may turn off the electricity to a home patient dependent upon the electricity to run their life-support machines. Remember that the return of any basic service to an offender should always be done with the intent of getting something from the offender, even if it is simple compliance with not following through on a threat to harm a hostage.

Any SWAT establishment that endeavors to resolve extremely violent crises must be fully trained, equipped, and prepared in all situations to properly apply overwhelming and deadly force.

The more your operators and officers train and participate in actual call-outs, the more accustomed and efficient they will be when they respond and resolve SWAT actions. The more they train, and the more careful the administering of discipline is, the chance of danger and use of force actually decreases.

The entire SWAT establishment must be sufficiently tried before they are led against an offender. This means they must all be well trained. It is much better to overcome an offender by fatigue, surprise, or intimidation than by general action, for in the latter instance luck may come into play, and in this instance fortune frequently has often a greater share than valor. The best plans are designed so that the offender is ignorant of the actions being taken by the SWAT establishment until the moment of execution. Opportunities in SWAT actions are often more to be depended on than courage.

Give offenders every reasonable opportunity to surrender peacefully and be sure to advertise this not only to them but also to bystanders and the media. This sincere encouragement helps to propel the offender toward your goal of not using force to take them into custody. It also builds confidence in bystanders through yielding numerous positive images of the teams attempting to peacefully resolve the crisis.

It is better to have a reserve of operators and officers than to place more than is needed in inner and outer perimeter duties.

A SWAT commander will rarely be unsuccessful if he knows the true abilities of his SWAT establishment and is also able to properly evaluate the offender's abilities and intentions. Smaller teams who rely upon valor and discipline are superior to larger teams that possess technology, resources, and numerous team members.

The nature of the critical terrain, avenues of approach, practiced decision making, and use of a disciplined SWAT establishment is often of more consequence than sheer courage or use of technology. Few are born brave; many become so through care and force of discipline. The SWAT establishment is strengthened by labor and enervated by idleness.

SWAT operators are not to be committed to resolve a static crisis unless confident of success.

A SWAT plan of action should concentrate on the use of novelty and surprise to throw and keep an offender in consternation; allowing an offender to have the initiative does nothing to resolve the crisis.

Those who were not prepared to pursue an offender frequently neglect perimeter needs, allowing an offender to breach a perimeter seems inclined to resign the arrest which could have been obtained.

A SWAT operator unsupplied with basic sustenance, obtainable equipment, or replacement supplies will eventually fail on an operation.

A SWAT team commander should always be prepared to maneuver call-outs to the static dimension, where all reasonable attempts at a peaceful resolution can be made. If less lethal uses of force are prudent, then those options should be explored next, with the last option in every instance being the use of lethal force. In some instances the use of lethal force will be the only option to safely resolved the crisis for everyone but the offender.

Initiative should be wrested from offenders at every opportunity and by all reasonable means. Every attempt and reasonable contrivance should be made to guide and manipulate the offender's actions to arrest.

Anytime the offender gains the initiative or is taking steps to gain initiative (such as actively barricading oneself), active steps of some type should be taken if possible to intervene, counter, and/or take the advantage away. Any effort to take initiative by an offender must be countered to place the SWAT operators back in a position of advantage, or otherwise place the SWAT establishment in immediate position to counter those steps. Distracting the offender or immediately emplacing breaching equipment and/or identifying alternate or multiple entry points is imperative.

Once offenders are located, the practice of space deprivation should be immediately considered. Space deprivation can be provided by using irritants to deny access to areas, physically occupying areas, and setting up ambuscades, or by using other physical means to deny offenders the ability to move about or otherwise gain intelligence on the teams operations and mission plans. Space deprivation is also a step toward imposing the will of the SWAT commander upon the offender. It limits the response options of the offender while increasing the options of the SWAT establishment to successfully resolve the crisis.

Any time the SWAT team must begin an operation in the dynamic mode, consideration should be given to turning the situation into a static mode.

Search warrant service must eventually be completed by team entry into a dwelling; however, after notification has been made, the structure can be breached and team members need enter only when it is safe. The use of breach and hold tactics, call-outs, canines, and remote vehicles if accessible, as well as gaining visual vantage points into the structure from outside *(rake and break) frequently significantly increase the SWAT operators safety margin.

The SWAT commander must depend on the discipline of the entire SWAT establishment. There is a famous quote that states, "Time is on your side." Remember the phrase, "Time is on your side," is not universal. "Time is on your side" only if the time is used wisely and productively, while simultaneously not allowing offenders to use that same time to their benefit. Unwise use of time allows the suspect to gain initiative and advantage. Thus, "Time is not on your side" when offenders use it to barricade, further injure hostages, plan escapes, gather intelligence on the actions being taken to induce their surrender, or make contingency plans against the SWAT operators.

The SWAT operators are in positions of great advantage when inner perimeter personnel are emplaced to not only perform perimeter duties but also potentially make an emergency or secondary entry into a structure or surprise a fleeing offender.

The SWAT commander cannot depend solely on either the numbers or courage of the SWAT establishment, and if obliged to engage every reasonable attempt should be made to distract the offender immediately prior to engaging (such as the use of distraction devices or techniques), while also bringing overwhelming force (economy of force) and numbers of operators on the offenders. Every attempt should be made to approach the offenders from multiple directions with the application of any force quickly and competently executed.

If the offender has taken advantage of some imminence or otherwise found means to barricade himself, all means should be taken to provide concealment, distractions, and instant breaching capabilities as necessary. Well-barricaded suspects in strong positions will most likely require additional crisis negotiation and the use of tiring techniques.

An offender should never be allowed to go mobile in a vehicle. Any vehicle the offender has access to should be disabled by some means, and any vehicle given to the offender should be provided by and specially prepared by SWAT establishment. The vehicle should be altered so that it can be electronically disabled and specially prepared to advantage of assault by SWAT operators if necessary.

The goal of every team should be to keep a dynamic

crisis that has turned static remaining static. Turning a static crisis dynamic should only be done when every reasonable advantage that can be gained has been gained. Additionally, the use of crisis negotiators and or a psychologist opinion as to the offenders possible actions should be carefully considered. Finally the decision on using dynamic means to resolve a crisis should be hinged on the order of safety priority, the public at large, officers, victims and hostages, and finally the suspect's personal safety being considered.

Deny the media access to all tactical plans and utilize a public information officer (PIO) to keep the media well informed only with information the SWAT commander and incident commander plans and wants released. A well-trained PIO will keep the media off of the incident commander's back and keep the media from interfering with the entire SWAT establishment as well as provide a positive image of the department. The PIO will accomplish this by giving the media lots of information, just nothing that will compromise the current plans or investigation. Enforce the boundaries of the command post and any staging areas, keep the media out of it and away from the general vicinity whenever possible.

The SWAT commander must ensure that alternative and contingency plans are made for every operation. If the offender knows where the operators or perimeter people are, move them. The offender should see nothing but what the commander wants him to see.

SWAT supervisors will tell SWAT commanders how they can resolve the operation and the steps they should take to resolve it. The SWAT commander should feel free to give advice based upon knowledge and experience. Together the commanders and supervisors should consult on the proper measures to be taken. After a plan of action has been confirmed, it should be communicated to the rest of the SWAT establishment. Special attention should be paid to make sure that the entire SWAT establishment is briefed upon their roles. Whenever possible, provide those taking actions opportunities to rehearse and practice their roles.

The SWAT establishment should be composed of the best officers and supervisors in the agency. They must all be highly motivated and talented officers. Remember that the officers will place themselves in harm's way. Make sure that they have been properly trained, disciplined, and equipped to do so. Recognize their efforts and ability to work in dangerous situations under stress. Thank them by ensuring they are well trained. Thank them by addressing and recognizing their actions individually and publicly. Finally, remember that a good commander will never commit a team into turning a static crisis into a dynamic crisis unless induced by opportunity or obliged by necessity.

Every SWAT commander should look for every opportunity to distress the offender and impose their will upon the offender more by wearing down the offender rather than by actual use of force. Taking a hostile offender into custody without injury during a crisis is generally considered a mark of consummate skill.

Many instructions might be given with regard to how the SWAT operators should physically perform their tasks. These skills and their mode of implementation frequently vary from department to department.

Every department contains experts in certain fields, even the simplest things such as how to hold a firearm are hotly debated between departments and experts. This branch of the art of SWAT has been brought to perfection by many different departments, all of which do things in different manners. Some modes are in fact more efficient than others. The SWAT establishment should be sending its members to other agencies of known excellence to learn the differing modes, disciplines, and tactics. Let it suffice that the establishment will then develop its own experts, and it will continually consider improvements to the drills and maneuvers, the arms, and the quality and management of the equipment. In short they will collect all that is useful from others and their mode of discipline will be sufficient.

The SWAT establishment should make every effort to conceal all dispositions for action and their means of action from offenders, lest they should counteract them and defeat the plans by proper expedients. It is a fact that the SWAT establishment should yield to the media's requests to view certain equipments. The media will also have numerous opportunities to view the SWAT establishment and its equipment in the field. However, all other contact and passing of information should be strictly controlled by the SWAT commander so as not to compromise operations. The media and public in general does not need to see the training or the rigor or violence of the training. Professionally minimize whenever possible actual media and public contact with the SWAT establishment and its equipment.

This abridgment on SWAT contains the maxims and instructions that have been formed and are generally approved by different agencies and confirmed by repeated experience. Before applying them, remember the media admires the skill in SWAT while alternately reporting or exploiting violent incidents to sell news. Criminals will endeavor in vain to imitate the SWAT establishment's dexterity in tactics. As long as strict discipline and a rigid doctrine toward training is maintained, organized crime groups and street gangs will never equal police individual skills with arms and team tactics. Even the United States military seeks out and utilizes police tactics and training in how to safely search areas, and to subdue and arrest combative persons. In conclusion, it is therefore a necessity that command staff have the qualifications to unite the science and art of SWAT and to convince the public that the agency's conduct and courage is being used for their safety.

PART 2.

DE RE MILITARI

INTRODUCTION/PREFACE

(This introduction was written for a British 1940 reprint of DE RE MILITARI. Author unknown.)

The most influential military treatise in the western world from Roman times to the 19th Century was Vegetius' DE RE MILITARI. Its impressions on our own traditions of discipline and organization are everywhere evident. The Austrian Field Marshal, Prince de Ligne, as late as 1770, called it a golden book and wrote: "A God, said Vegetius, inspired the legion, but for myself, I find that a God inspired Vegetius." Richard Coeur de Lion carried DE RE MILITARI everywhere with him in his campaigns, as did his father, Henry II of England. Around 1000 A. D. Vegetius was the favorite author of Foulques the Black, the able and ferocious Count of Anjou. Numerous manuscript copies of Vegetius circulated in the time of Charlemagne and one of them was considered a necessity of life by his commanders. A manuscript Vegetius was listed in the will of Count Everard de Frejus, about 837 A. D., in the time of Ludwig the Just.

In his Memoirs, Montecuculli, the conqueror of the Turks at St. Gotthard, wrote: "However, there are spirits bold enough to believe themselves great captains as soon as they know how to handle a horse, carry a lance at charge in a tournament, or as soon as they have read the precepts of Vegetius." Such was the reputation of Vegetius for a thousand years. Manuscript copies dating from the 10th to the 15th centuries are extant to the number of 150. DE RE MILITARI was translated into English, French, and Bulgarian before the invention of printing. The first printed edition was made in Utrecht in 1473. It was followed in quick succession by editions in Cologne, Paris and Rome. It was first published in English by Caxton, from an English manuscript copy, in 1489.

Flavius Vegetius Renatus was a Roman of high rank. In some manuscripts he is given the title of count. Raphael of Volterra calls him a Count of Constantinople. Little is known of his life. It is apparent from his book that he had not had extensive practical experience as a soldier. . He states quite frankly that his purpose was to collect and synthesize from ancient manuscripts and regulations the military customs and wisdom that made ancient Rome great. According to his statement, his principal sources were Cato the Elder, Cornelius Celsus, Paternus, Frontinus, and the regulations and ordinances of Augustus, Trajan and Hadrian.

The Emperor Valentinian, to whom the book is dedicated, is believed to be the second emperor of that name. He evidently was not Valentinian I since his successor, Gratian, is named in the book. Between the reign of Valentinian II and Valentinian III, Rome was taken and burned by Alaric, King of the Goths, an event that unquestionably would have been mentioned had it occurred before the book was written. Vegetius mentions the defeat of the Roman armies by the Goths, but probably refers to the battle of Adrianople where Valens, the colleague of Valentinian I, was killed.

It is a paradox that DE RE MILITARI, which was to become a military bible for innumerable generations of European soldiers, was little used by the Romans for whom it was written. The decay of the Roman armies had progressed too far to be arrested by

Vegetius' pleas for a return to the virtues of discipline and courage of the ancients. At the same time Vegetius' hope for a revival of the ancient organization of the legion was impracticable. Cavalry had adopted the armor of the foot soldier and was just commencing to become the principal arm of the military forces. The heavy armed foot-soldier, formerly the backbone of the legion, was falling a victim of his own weight and immobility, and the light-armed infantry, unable to resist the shock of cavalry, was turning more and more to missile weapons. By one of the strange mutations of history, when later the cross-bow and gun-powder deprived cavalry of its shock-power, the tactics of Vegetius again became ideal for armies, as they had been in the times from which he drew his inspiration.

Vegetius unceasingly emphasized the importance of constant drill and severe discipline and this aspect of his work was very tiresome to the soldiers of the middle ages, the feudal system lending itself but poorly to discipline. "Victory in war," he states in his opening sentence, "does not depend entirely upon numbers or mere courage; only skill and discipline will insure it." His first book is devoted to the selection, training and discipline of recruits. He insists upon the utmost meticulousness in drill. "No part of drill is more essential in action than for soldiers to keep their ranks with the greatest exactness." His description of the many arms which the Roman soldier was required to become expert in reminds one of the almost innumerable duties of the present day infantryman. Recruits were to be hardened so as to be able to march twenty miles in half a summer's day at ordinary step and twenty-four miles at quick step.

It was the ancient regulation that practice marches of this distance must be made three times a month.

The second book deals with the organization and officers of the legion, the ancient system of promotion, and how to form the legion for battle. We find the Romans provided for soldier's deposits, just as is done in the American army today; that guard and duty rosters were kept in those days as now; and that the Roman system of guard duty is only slightly different from our manual for interior guard duty. The field music is described and is an ornamental progenitor of that in use in United States. The legion owed its success, according to Vegetius, to its arms and its machines, as well as to the bravery of its soldiers. The legion had fifty-five ballista for throwing darts and ten onagri, drawn by oxen, for throwing stones. Every legion carried its ponton equipment, "small boats hollowed out of a single piece of cimber, with long cables or chains to fasten them together." And in addition were "whatever is necessary for every kind of service, that the encampments may have all the strength and conveniences of a fortified city." Trains of workmen were provided to perform all the duties now performed by the various services in armies.

The third book deals with tactics and strategy and it was this portion of Vegetius that influenced war in the Middle Ages so greatly. He explains the use of reserves, attributing this invention to the Spartans, from whom the Romans adopted it. "It is much better to have several bodies of reserves than to extend your front too much" - an injunction as good today as when it was written. Encircling pursuit is described. The terrain is not overlooked. "The nature of the ground is often of more consequence than courage." The enemy should be estimated carefully. "It is essential to know the character of the enemy and of their principal officers-whether they be rash or cautious, enterprising or timid, whether they fight from careful calculation or from chance."

Vegetius' work is filled with maxims that have become a part of our everyday life. "He, therefore, who aspires to peace should prepare for war." "The ancients preferred discipline to numbers." "In the midst of peace, war is looked upon as an object too distant to merit consideration." "Few men are born brave; many become so through training and force of discipline." Vegetius was a reformer who attempted to restore the degenerate Romans of the 4th Century to the military virtues of the ancients, whom he never ceases to laud. His little book was made short and easy to read, so as not to frighten, by a too arduous text, the readers whom he hoped to convince. He constantly gives the example of the " Ancients" to his contemporaries. The result is a sort of perfume of actuality, which had much to do with his success. It still is interesting reading and still is the subject of modern commentaries. No less than forty have appeared in Germany in the 19th and 20th centuries. Revue Mititare Generate (France) and our own Infantry Journal carried articles on Vegetius in 1938. Dankfried Schenk published an interesting article in Klio in 1930, which gives Vegetius the highest place among the writers of his time.

The present edition includes the first three books of Vegetius' work, omitting only repetitions. The fourth and fifth books, both very brief, deal with the attack and defense of fortified places and with naval operations. These are of interest only to military antiquarians and for that reason have not been included.

The present translation was made by Lieutenant John Clarke and published in London in 1767. It is the best available in English and has been edited only to the minimum extent necessary to conform to modern usage.

An excellent discussion of Vegetius can be found in Warfare, by Spaulding, Nickerson and Wright, page 294, et sequens, Harcourt Brace & Co., 1925. Delpech, La Tactique au 13me Siecte, Paris, 1886, gives the best account of the influence of Vegetius on European military thought. Hans Delbruck's discussion of Vegetius in Geschichte der Kriegskunft, Vol. II, Berlin, 1921, although brief, is very acute.

4

THE SELECTION AND TRAINING OF NEW LEVIES

To the Emperor Valentinian:

It has been an old custom for authors to offer to their Princes the fruits of their studies in belles letters, from a persuasion that no work can be published with propriety but under the auspices of the Emperor, and that the knowledge of a Prince should be more general, and of the most important kind, as its influence is felt so keenly by all his subjects. We have many instances of the favorable reception which Augustus and his illustrious successors conferred on the works presented to them; and this encouragement of the Sovereign made the sciences flourish. The consideration of Your Majesty's superior indulgence for attempts of this sort, induced me to follow this example, and makes me at the same time almost forget my own inability when compared with the ancient writers. One advantage, however, I derive from the nature of this work, as it requires no elegance of expression, or extraordinary share of genius, but only great care and fidelity in collecting and explaining, for public use, the instructions and observations of our old historians of military affairs, or those who wrote expressly concerning them.

My design in this treatise is to exhibit in some order the peculiar customs and usages of the ancients in the choice and discipline of their new levies. Nor do I presume to offer this work to Your Majesty from a supposition that you are not acquainted with every part of its contents; but that you may see that the same salutary dispositions and regulations which your own wisdom prompts You to establish for the happiness of the Empire, were formerly observed by the founders thereof; and that Your Majesty may find with ease in this abridgement whatever is most useful on so necessary and important a subject.

The Roman Discipline the Cause of Their Greatness

Victory in war does not depend entirely upon numbers or mere courage; only skill and discipline will insure it. We find that the Romans owed the conquest of the world to no other cause than continual military training, exact observance of discipline in their camps and unwearied cultivation of the other arts of war. Without these, what chance would the inconsiderable numbers of the Roman armies have had against the multitudes of the Gauls? Or with what success would their small size have been opposed to the prodigious stature of the Germans? The Spaniards surpassed us not only in numbers, but in physical strength. We were always inferior to the Africans in wealth and unequal to them in deception and stratagem. And the Greeks, indisputably, were far superior to us in skill in arts and all kinds of knowledge.

But to all these advantages the Romans opposed unusual care in the choice of their levies and in their military training. They thoroughly understood the importance of hardening them by continual practice, and of training them to every maneuver that might happen in the line and in action. Nor were they less strict in punishing idleness and sloth. The courage of a soldier is heightened by his knowledge of his profession, and he only wants an opportunity to execute what he is convinced he has been perfectly taught. A handful of men, inured to war, proceed to certain victory, while on the contrary numerous armies of raw and undisciplined troops are but multitudes of men dragged to slaughter.

The Selection of Recruits

To treat our subject with some method, we shall first examine what provinces or nations are to be preferred for supplying the armies with recruits. It is certain that every country produces both brave men and cowards; but it is equally as certain that some nations are naturally more warlike than others, and that courage, as well as strength of body, depends greatly upon the influence of the different climates.

We shall next examine whether the city or the country produces the best and most capable soldiers. No one, I imagine, can doubt that the peasants are the most fit to carry arms for they from their infancy have been exposed to all kinds of weather and have been brought up to the hardest labor. They are able to endure the greatest heat of the sun, are unacquainted with the use of baths, and are strangers to the other luxuries of life. They are simple, content with little, inured to all kinds of fatigue, and prepared in some measure for a military life by their continual employment in their country-work, in handling the spade, digging trenches and carrying burdens. In cases of necessity, however, they are sometimes obliged to make levies in the cities. And these men, as soon as enlisted, should be taught to work on entrenchments, to march in ranks, to carry heavy burdens, and to bear the sun and dust. Their meals should be coarse and moderate; they should be accustomed to lie sometimes in the open air and sometimes in tents. After this, they should be instructed in the use of their arms. And if any long expedition is planned, they should be encamped as far as possible from the temptations of the city. By these precautions their minds, as well as their bodies, will properly be prepared for the service.

I realize that in the first ages of the Republic, the Romans always raised their armies in the city itself, but this was at a time when there were no pleasures, no luxuries to enervate them. The Tiber was then their only bath, and in it they refreshed themselves after their exercises and fatigues in the field by swimming. In those days the same man was both soldier and farmer, but a farmer who, when occasion arose, laid aside his tools and put on the sword. The truth of this is confirmed by the instance of Quintius Cincinnatus, who was following the plow when they came to offer him the dictatorship. The chief strength of our armies, then, should be recruited from the country. For it is certain that the less a man is acquainted with the sweets of life, the less reason he has to be afraid of death.

The Proper Age for Recruits

If we follow the ancient practice, the proper time for enlisting youth into the army is at their entrance into the age of puberty. At this time instructions of every kind are more quickly imbibed and more lastingly imprinted on the mind. Besides this, the indispensable military exercises of running and leaping must be acquired before the limbs are too much stiffened by age. For it is activity, improved by continual practice, which forms the useful and good soldier. Formerly, says Sallust, the Roman youth, as soon as they were of an age to carry arms, were trained in the Strictest manner in their camps to all the fatigues and exercises of war. For it is certainly better that a soldier, perfectly disciplined, should, through emulation, repine at his not being yet arrived at a proper age for action, than have the mortification of knowing it is past. A sufficient time is also required for his instruction in the different branches of the service. It is no easy matter to train the horse or foot archer, or to form the legionary soldier to every part of the drill, to teach him not to quit his post, to keep ranks, to take a proper aim and throw his missile weapons with force, to dig trenches, to plant palisades, how to manage his shield, glance off the blows of the enemy, and how to parry a stroke with dexterity. A soldier, thus perfect in his business, so far from showing any backwardness to engage, will be eager for an opportunity of signaling himself.

Their Size

We find the ancients very fond of procuring the tallest men they could for the service, since the standard for the cavalry of the wings and for the infantry of the first legionary cohorts was fixed at six feet, or at least five feet ten inches. These requirements might easily be kept up in those times when such numbers followed the profession of arms and before it was the fashion for the flower of Roman youth to devote themselves to the civil offices of state. But when necessity requires it, the height of a man is not to be regarded so much as his strength; and for this we have the authority of Homer, who tells us that the deficiency of stature in Tydeus was amply compensated by his vigor and courage.

Signs of Desirable Qualities

Those employed to superintend new levies should be particularly careful in examining the features of their faces, their eyes, and the make of their limbs, to enable them to form a true judgment and choose such as are most likely to prove good soldiers. For experience assures us that there are in men, as well as in horses and dogs, certain signs by which their virtues may be discovered. The young soldier, therefore, ought to have a lively eye, should carry his head erect, his chest should be broad, his shoulders muscular and brawny, his fingers long, his arms strong, his waist small, his shape easy, his legs and feet rather nervous than fleshy. When all these marks are found in a recruit, a little height may be dispensed with, since it is of much more importance that a soldier should be strong than tall.

Trades Proper for New Levies

In choosing recruits regard should be given to their trade. Fishermen, fowlers, confectioners, weavers, and in general all whose professions more properly belong to women should, in my opinion, by no means be admitted into the service. On the contrary, smiths, carpenters, butchers, and huntsmen are the most proper to be taken into it. On the careful choice of soldiers depends the welfare of the Republic, and the very essence of the Roman Empire and its power is so inseparably connected with this charge, that it is of the highest importance not to be intrusted indiscriminately, but only to persons whose fidelity can be relied on. The ancients considered Sertorius' care in this point as one of the most eminent of his military qualifications. The soldiery to whom the defense of the Empire is consigned and in whose hands is the fortune of war, should, if possible, be of reputable families and unexceptionable in their manners. Such sentiments as may be expected in these men will make good soldiers. A sense of honor, by preventing them from behaving ill, will make them victorious.

But what good can be expected from a man by nature a coward, though ever so well disciplined or though he has served ever so many campaigns? An army raised without proper regard to the choice of its recruits was never yet made good by length of time; and we are now convinced by fatal experience that this is the source of all our misfortunes. So many defeats can only be imputed to the effects of a long peace which has made us negligent and careless in the choice of our levies and to the inclination so prevalent among the better sort in preferring the civil posts of the government to the profession of arms and to the shameful conduct of the superintendents, who, through interest or connivance, accept many men which those who are obliged to furnish substitutes for the army choose to send, and admit such men into the service as the masters themselves would not even keep for servants. Thus it appears that a trust of such importance should be committed to none but men of merit and integrity.

The Military Mark

The recruit, however, should not receive the military mark* as soon as enlisted. He must first be tried if fit for service; whether he has sufficient activity and strength; if he has capacity to learn his duty; and whether he has the proper degree of military courage. For many, though promising enough in appearance, are found very unfit upon trial. These are to be rejected and replaced by better men; for it is not numbers, but bravery which carries the day.

After their examination, the recruits should then receive the military mark, and be taught the use of their arms by constant and daily exercise. But this essential custom has been abolished by the relaxation introduced by a long peace. We cannot now expect to find a man to teach what he never learned himself. The only method, therefore, that remains of recovering the ancient customs is by books, and by consulting the old historians. But they are of little service to us in this respect, as they only relate the exploits and events of wars, and take no notice of the objects of our present enquiries, which they considered as universally known.

Initial Training

The first thing the soldiers are to be taught is the military step, which can only be acquired by constant practice of marching quick and together. Nor is anything of more consequence either on the march or in the line than that they should keep their ranks with the greatest exactness. For troops who march in an irregular and disorderly manner are always in great danger of being defeated. They should march with the common military step twenty miles in five summer-hours, and with the full step, which is quicker, twenty-four miles in the same number of hours. If they exceed this pace, they no longer march but run, and no certain rate can be assigned.

But the young recruits in particular must be exercised in running, in order to charge the enemy with great vigor; occupy, on occasion, an advantageous post with greater expedition, and prevent the enemy in their designs upon the same; that they may, when sent to reconnoiter, advance with speed, return with greater celerity and more easily come up with the enemy in a pursuit.

Leaping is another very necessary exercise, to enable them to pass ditches or embarrassing eminences of any kind without trouble or difficulty. There is also another very material advantage to be derived from these exercises in time of action; for a soldier who advances with his javelin, running and leaping, dazzles the eyes of his adversary, strikes him with terror, and gives him the fatal stroke before he has time to put himself on his defense. Sallust, speaking of the excellence of Pompey the Great in these particulars, tells us that he disputed the superiority in leaping with the most active, in running with the most swift, and in exercises of strength with the most robust. Nor would he ever have been able to have opposed Serrorius with success, if he had not prepared both himself and his soldiers for action by continual exercises of this sort.

To Learn to Swim

Every young soldier, without exception, should in the summer months be taught to swim; for it is sometimes impossible to pass rivers on bridges, but the flying and pursuing army both are often obliged to swim over them. A sudden melting of snow or fall of rain often makes them overflow their banks, and in such a situation, the danger is as great from ignorance in swimming as from the enemy. The ancient Romans, therefore, perfected in every branch of the military art by a continued series of wars and perils, chose the Field of Mars as the most commodious for their exercises on account of its vicinity to the Tiber, that the youth might therein wash off the sweat and dust, and refresh themselves after their fatigues by swimming. The cavalry also as well as the infantry, and even the horses and the servants of the army should be accustomed to this exercise, as they are all equally liable to the same accidents.

The Post Exercise

We are informed by the writings of the ancients that, among their other exercises, they had that of the post. They gave their recruits round bucklers woven with willows, twice as heavy as those used on real service, and wooden swords double the weight of the common ones. They exercised them with these at the post both morning and afternoon.

This is an invention of the greatest use, not only to soldiers, but also to gladiators. No man of either profession ever distinguished himself in the circus or field of battle, who was not perfect in this kind of exercise. Every soldier, therefore, fixed a post firmly in the ground, about the height of six feet. Against this, as against a real enemy, the recruit was exercised with the above mentioned arms, as it were with the common shield and sword, sometimes aiming At the head or face, sometimes at the sides, at others endeavoring to strike at the thighs or legs. He was instructed in what manner to advance and retire, and in short how to take every advantage of his adversary; but was thus above all particularly cautioned not to lay himself open to his antagonist while aiming his stroke at him.

Not to Cut, But to Thrust with the Sword

They were likewise taught not to cut but to thrust with their swords. For the Romans not only made a jest of those who fought with the edge of that weapon, but always found them an easy conquest. A stroke with the edges, though made with ever so much force, seldom kills, as the vital parts of the body are defended both by the bones and armor. On the contrary, a stab, though it penetrates but two inches, is generally fatal. Besides in the attitude of striking, it is impossible to avoid exposing the right arm and side; but on the other hand, the body is covered while a thrust is given, and the adversary receives the point before he sees the sword. This was the method of fighting principally used by the Romans, and their reason for exercising recruits with arms of such a weight at first was, that when they came to carry the common ones so much lighter, the greater difference might enable them to act with greater security and alacrity in time of action.

The Drill Called Armatura

The new levies also should be taught by the masters at arms the system of drill called armatura, as it is still partly kept up among us. Experience even at this time convinces us that soldiers, perfect therein, are of the most service in engagements. And they afford certain proofs of the importance and effects of discipline in the difference we see between those properly trained in this branch of drill and the other troops. The old Romans were so conscious of its usefulness that they rewarded the masters at arms with a double allowance of provision. The soldiers who were backward in this drill were punished by having their allowance in barley. Nor did they receive it as usual, in wheat, until they had, in the presence of the prefect, tribunes, or other principal officers of the legion, showed sufficient proofs of their knowledge of every part of their study.

No state can either be happy or secure that is remiss and negligent in the discipline of its troops. For it is not profusion of riches or excess of luxury that can influence our enemies to court or respect us. This can only be effected by the terror of our arms. It is an observation of Cato that. misconduct in the common affairs of life may be retrieved, but that it is quite otherwise in war, where errors are fatal and without remedy, and are followed by immediate punishment. For the consequences of engaging an enemy, without skill or courage, is that part of the army is left on the field of battle, and those who remain receive such an impression from their defeat that they dare not afterwards look the enemy in the face.

The Use of Missile Weapons

Besides the aforementioned exercise of the recruits at the post, they were furnished with javelins of greater weight than common, which they were taught to throw at the same post. And the masters at arms were very careful to instruct them how to cast them with a proper aim and force. This practice strengthens the arm and makes the soldier a good marksman.

The Use of the Bow

A third or fourth of the youngest and fittest soldiers should also be exercised at the post with bows and arrows made for that purpose only. The masters for this branch must be chosen with care and must apply themselves diligently to teach the men to hold the bow in a proper position, to bend it with strength, to keep the left hand steady. to draw the right with skill, to direct both the attention and the eye to the object, and to take their aim with equal certainty either on foot or on horseback. But this is not to be acquired without great application, nor to be retained without daily exercise and practice.

The utility of good archers in action is evidently demonstrated by Cato in his treatise on military discipline. To the institution of a body of troops of this sort Claudius owed his victory over an enemy who, till that time, had constantly been superior to him. Scipio Africanus, before his battle with the Numantines, who had made a Roman army ignominiously pass under the yoke, thought he could have no likelihood of success except by mingling a number of select archers with every century.

The Sling

Recruits are to be taught the art of throwing stones both with the hand and sling. The inhabitants of the Balearic Islands are said to have been the inventors of slings, and to have managed them with surprising dexterity, owing to the manner of bringing up their children. The children were not allowed to have their food by their mothers till they had first struck it with their sling. Soldiers, notwithstanding their defensive armor, are often more annoyed by the round stones from the sling than by all the arrows of the enemy. Stones kill without mangling the body, and the contusion is mortal without loss of blood. It is universally known the ancients employed slingers in all their engagements. There is the greater reason for instructing all troops, without exception, in this exercise, as the sling cannot be reckoned any incumbrance, and often is of the greatest service, especially when they are obliged to engage in stony places, to defend a mountain or an eminence, or to repulse an enemy at the attack of a castle or city.

The Loaded Javelin

The exercise of the loaded javelins, called martiobarbuli, must not be omitted. We formerly had two legions in lllyricum, consisting of six thousand men each, which from their extraordinary dexterity and skill in the use of these weapons were distinguished by the same appellation. They supported for a long time the weight of all the wars and distinguished themselves so remarkably that the Emperors Diocletian and Maximian on their accession honored them with the titles of Jovian and Herculean and preferred them before all the other legions. Every soldier carries five of these javelins in the hollow of his shield. And thus the legionary soldiers seem to supply the place of archers, for they wound both the men and horses of the enemy before they come within reach of the common missile weapons.

To Be Taught to Vault

The ancients strictly obliged both the veteran soldiers and recruits to a constant practice of vaulting. It has indeed reached our cimes, although little regard is paid to it at present. They had wooden horses for that purpose placed in winter under cover and in summer in the field. The young soldiers were taught to vault on them at first without arms, afterwards completely armed. And such was their attention to this exercise that they were accustomed to mount and dismount on either side indifferently, with their drawn swords or lances in their hands. By assiduous practice in the leisure of peace, their cavalry was brought to such perfection of discipline that they mounted their horses in an instant even amidst the confusion of sudden and unexpected alarms.

And to Carry Burdens

To accustom soldiers to carry burdens is also an essential part of discipline. Recruits in particular should be obliged frequently to carry a weight of not less than sixty pounds (exclusive of their arms), and to march with it in the ranks. This is because on difficult expeditions they often find themselves under the necessity of carrying their provisions as well as their arms. Nor will they find this troublesome when inured to it by custom, which makes everything easy. Our troops in ancient times were a proof of this, and Virgil has remarked it in the following lines:

> The Roman soldiers, bred in war's alarms,
> Bending with unjust loads and heavy arms,
> Cheerful their toilsome marches undergo,
> And pitch their sudden camp before the foe.

The Arms of the Ancients

The manner of arming the troops comes next under consideration. But the method of the ancients no longer is followed. For though after the example of the Goths, the Alans and the Huns, we have made some improvements in the arms of the cavalry, yet it is plain the infantry are entirely defenseless. From the foundation of the city till the reign of the Emperor Gratian, the foot wore cuirasses and helmets. But negligence and sloth having by degrees introduced a total relaxation of discipline, the soldiers began to think their armor too heavy, as they seldom put it on. They first requested leave from the Emperor to lay aside the cuirass and afterwards the helmet. In consequence of this, our troops in their engagements with the Goths were often overwhelmed with their showers of arrows. Nor was the necessity of obliging the infantry to resume their cuirasses and helmets discovered, notwithstanding such repeated defeats, which brought on the destruction of so many great cities.

Troops, defenseless and exposed to all the weapons of the enemy, are more disposed to fly than fight. What can be expected from a foot-archer without cuirass or helmet, who cannot hold at once his bow and shield; or from the ensigns whose bodies are naked, and who cannot at the same time carry a shield and the colors? The foot soldier finds the weight of a cuirass and even of a helmet intolerable. This is because he is so seldom exercised and rarely puts them on.

But the case would be quite different, were they even heavier than they are, if by constant practice he had been accustomed to wear them. But it seems these very men, who cannot support the weight of the ancient armor, think nothing of exposing themselves without defense to wounds and death, or, which is worse, to the shame of being made prisoners, or of betraying their country by flight; and thus to avoid an inconsiderable share of exercise and fatigue, suffer themselves ignominiously to be cut in pieces. With what propriety could the ancients call the infantry a wall, but that in some measure they resembled it by the complete armor of the legionary soldiers who had shields, helmets, cuirasses, and greaves of iron on the right leg; and the archers who had gauntlets on the left arm. These were the defensive arms of the legionary soldiers. Those who fought in the first line of their respective legions were called principes, in the second hastati, and in third triarii.

The triarii, according to their method of discipline, rested in time of action on one knee, under cover of their shields, so that in this position they might be less exposed to the darts of the enemy than if they stood upright; and also, when there was a necessity for bringing them up, that they might be fresh, in full vigor and charge with the greater impetuosity. There have been many instances of their gaining a complete victory after the entire defeat of both the principes and hastati.

The ancients had likewise a body of light infantry, slingers, and ferentarii (the light troops), who were generally posted on the wings and began the engagement. The most active and best disciplined men were selected for this service; and as their number was not very great, they easily retired in case of a repulse through the intervals of the legion, without thus occasioning the least disorder in the line.

The Pamonian leather caps worn by our soldiers were formerly introduced with a different design. The ancients obliged the men to wear them at all times so that being constantly accustomed to have the head covered they might be less sensible of the weight of the helmet.

As to the missile weapons of the infantry, they were javelins headed with a triangular sharp iron, eleven inches or a foot long, and were called piles. When once fixed in the shield it was impossible to draw them out, and when thrown with force and skill, they penetrated the cuirass without difficulty. At present they are seldom used by us, but are the principal weapon of the barbarian heavy-armed foot. They are called bebrae, and every man carries two or three of them to battle.

It must be observed that when the soldiers engage with the javelin, the left foot should be advanced, for, by this attitude the force required to throw it is considerably increased. On the contrary, when they are close enough to use their piles and swords, the right foot should be advanced, so that the body may present less aim to the enemy, and the right arm be nearer and in a more advantageous position for striking. Hence it appears that it is as necessary to provide soldiers with defensive arms of every kind as to instruct them in the use of offensive ones. For it is certain a man will fight with greater courage and confidence when he finds himself properly armed for defense.

Entrenched Camps

Recruits are to be instructed in the manner of entrenching camps, there being no part of discipline so necessary and useful as this. For in a camp, well chosen and entrenched, the troops both day and night lie secure within their works, even though in view of the enemy. It seems to resemble a fortified city which they can build for their safety wherever they please. But this valuable art is now entirely lost, for it is long since any of our camps have been fortified either with trenches or palisades. By this neglect our forces have been often surprised by day and night by the enemy's cavalry and suffered very severe losses. The importance of this custom appears not only from the danger to which troops are perpetually exposed who encamp without such precautions, but from the distressful situation of an army that, after receiving a check in the field, finds itself without retreat and consequently at the mercy of the enemy. A camp, especially in the neighborhood of an enemy, must be chosen with great care. Its situation should be strong by nature, and there should be plenty of wood, forage and water. If the army is to continue in it any considerable time, attention must be had to the salubrity of the place. The camp must not be commanded by any higher grounds from whence it might be insulted or annoyed by the enemy, nor must the location be liable to floods which would expose the army to great danger. The dimensions of the camps must be determined by the number of troops and quantity of baggage, that a large army may have room enough, and that a small one may not be obliged to extend itself beyond its proper ground. The form of the camps must be determined by the site of the country, in conformity to which they must be square, triangular or oval. The Praetorian gate should either front the east or the enemy. In a temporary camp it should face the route by which the army is to march. Within this gate the tents of the first centuries or cohorts are pitched, and the dragons* and other ensigns planted. The Decumane gate is directly opposite to the Praetorian in the rear of the camp, and through this the soldiers are conducted to the place appointed for punishment or execution.

There are two methods of entrenching a camp. When the danger is not imminent, they carry a slight ditch round the whole circuit, only nine feet broad and seven deep. With the turf taken from this they make a kind of wall or breastwork three feet high on the inner side of the ditch. But where there is reason to be apprehensive of attempts of the enemy, the camp must be surrounded with a regular ditch twelve feet broad and nine feet deep perpendicular from the surface of the ground. A parapet is then raised on the side next the camp, of the height of four feet, with hurdles and fascines properly covered and secured by the earth taken out of the ditch. From these dimensions the interior height of the intrenchment will be found to be thirteen feet, and the breadth of the ditch twelve. On the top of the whole are planted strong palisades which the soldiers carry constantly with them for this purpose. A sufficient number of spades, pickaxes, wicker baskets and tools of all kinds are to be provided for these works.

There is no difficulty in carrying on the fortifications of a camp when no enemy is in sight. But if the enemy is near, all the cavalry and half the infantry are to be drawn up in order of battle to cover the rest of the troops at work on the entrenchments and be ready to receive the enemy if they offer to attack. The centuries are employed by turns on the work and are regularly called to the relief by a crier till the whole is completed. It is then inspected and measured by the centurions, who punish such as have been indolent or negligent. This is a very important point in the discipline of young soldiers, who when properly trained to it will be able in an emergency to fortify their camp with skill and expedition.

Evolutions

No part of drill is more essential in action than for soldiers to keep their ranks with the greatest exactness, without opening or closing too much. Troops too much crowded can never fight as they ought, and only embarrass one another. If their order is too open and loose, they give the enemy an opportunity of penetrating. Whenever this happens and they are attacked in the rear, universal disorder and confusion are inevitable. Recruits should therefore be constantly in the field, drawn up by the roll and formed at first into a single rank. They should learn to dress in a straight line and to keep an equal and just distance between man and man. They must then be ordered to double the rank, which they must perform very quickly, and instantly cover their file leaders. In the next place, they are to double again and form four deep. And then the triangle or, as it is commonly called, the wedge, a disposition found very serviceable in action. They must be taught to form the circle or orb; for well-disciplined troops, after being broken by the enemy, have thrown themselves into this position and have thereby prevented the total rout of the army. These evolutions, often practiced in the field of exercise, will be found easy in execution on actual service.

Monthly Marches

It was a constant custom among the old Romans, confirmed by the Ordinances of Augustus and Hadrian, to exercise both cavalry and infantry three times in a month by marches of a certain length. The foot were obliged to march completely armed the distance of ten miles from the camp and return, in the most exact order and with the military step which they changed and quickened on some part of the march. Their cavalry likewise, in troops and properly armed, performed the same marches and were exercised at the same time in their peculiar movement and evolutions; sometimes, as if pursuing the enemy, sometimes retreating and returning again with greater impetuosity to the charge. They made these marches not in plain and even ground only, but both cavalry and infantry were ordered into difficult and uneven places and to ascend or descend mountains, to prepare them for all kinds of accidents and familiarize them with the different maneuvers that the various situations of a country may require.

Conclusion

These military maxims and instructions, invincible Emperor, as a proof of my devotion and zeal for your service, I have carefully collected from the works of all the ancient authors on the subject. My design herein is to point out the certain method of forming good and serviceable armies, which can only be accomplished by an exact imitation of the ancients in their care in the choice and discipline of their levies. Men are not degenerated in point of courage, nor are the countries that produced the Lacedaemonians, the Athenians, the Marsians, the Samnites, the Peligni and even the Romans themselves, yet exhausted. Did not the Epirots acquire in former times a great reputation in war? Did not the Macedonians and Thessalians, after conquering the Persians, penetrate even into India? And it is well known that the warlike dispositions of the Dacians, Moesians and Thracians gave rise to the fable that Mars was born among them.

To pretend to enumerate the different nations so formidable of old, all which now are subject to the Romans, would be tedious. But the security established by long peace has altered their dispositions, drawn them off from military to civil pursuits and infused into them a love of idleness and ease. Hence a relaxation of military discipline insensibly ensued, then a neglect of it, and it sunk at last into entire oblivion. Now will it appear surprising that this alteration should have happened in latter times, if we consider that the peace, which lasted about twenty years or somewhat more after the first Punic war, enervated the Romans, before everywhere victorious, by idleness and neglect of discipline to such a degree, that in the second Punic war they were not able to keep the field against Hannibal. At last, after the defeat of many consuls and the loss of many officers and armies, they were convinced that the revival of discipline was the only road to victory and thereby recovered their superiority. The necessity, therefore, of discipline cannot be too often inculcated, as well as the strict attention requisite in the choice and training of new levies. It is also certain that it is a much less expense to a State to train its own subjects to arms than to take foreigners into its pay.

* This mark was imprinted on the hands of the soldiers, either with a hot iron, or in some other manner. It was indelible.

* The dragon was the particular ensign of each cohort.

5

THE ORGANIZATION OF THE LEGION

To the Emperor Valentinian:

Such a continued series of victories and triumphs proved incontestably Your Majesty's full and perfect knowledge of the military discipline of the ancients. Success in any profession is the most certain mark of skill in it. By a greatness of mind, above human comprehension Your Majesty condescends to seek instruction from the ancients, notwithstanding your own recent exploits surpass antiquity itself. On receiving Your Majesty's orders to continue this abridgement, not so much for your instruction as convenience, I knew not how to reconcile my devotion to Your commands with the respect due to Your Majesty. Would it not be the greatest height of presumption to pretend to mention the art of war to the Lord and Master of the world and the Conqueror of all the barbarous nations, unless it were to describe his own actions? But disobedience to the will of so great a Prince would be both highly criminal and dangerous. My obedience, therefore, made me presumptuous, from the apprehensions of appearing more so by a contrary conduct. And in this I was not a little encouraged by the late instance of Your Majesty's indulgence. My treatise on the choice and discipline of new levies met with a favorable reception from Your Majesty, and since a work succeeded so well, composed of my own accord, I can have no fears for one undertaken by your own express commands.

The Military Establishment

The military establishment consists of three parts, the cavalry, infantry and marine. The wings of cavalry were so called from their similitude to wings in their extension on both sides of the main body for its protection. They are now called vexillations from the kind of standards peculiar to them. The legionary horse are bodies particularly annexed to each legion, and of a different kind; and on their model were organized the cavalry called Ocreati, from the light boots they wear. The fleet consists of two divisions, the one of men of war called Liburnae, and the other of armed sloops. The cavalry are designed for plains. Fleets are employed for the protection of seas and rivers. The infantry are proper for the defense of eminences, for the garrisons of cities and are equally serviceable in plain and in uneven ground. The latter, therefore, from their facility of acting everywhere, are certainly the most useful and necessary troops to a state exclusively of the consideration of their being maintained at a less expense. The infantry are divided into two corps, the legions and auxiliaries, the latter of which are furnished by allies or confederates. The peculiar strength of the Romans always consisted in the excellent organization of their legions. They were so denominated ab eligendo, from the care and exactness used in the choice of the soldiers. The number of legionary troops in an army is generally much more considerable than that of the auxiliaries.

Difference Between the Legions and Auxiliaries

The Macedonians, the Greeks and the Dardanians formed their troops into phalanxes of eight thousand men each. The Gauls, Celtiberians and many other barbarous nations divided their armies into bodies of six thousand each. The Romans have their legions usually six thousand strong, sometimes more.

We shall now explain the difference between the legions and the auxiliaries. The latter are hired corps of foreigners assembled from different parts of the Empire, made up of different numbers, without knowledge of one another or any tie of affection. Each nation has its own peculiar discipline, customs and manner of fighting. Little can be expected from forces so dissimilar in every respect, since it is one of the most essential points in military undertakings that the whole army should be put in motion and governed by one and the same order. But it is almost impossible for men to act in concert under such varying and unsettled circumstances. They are, however, when properly trained and disciplined, of material service and are always joined as light troops with the legions in the line. And though the legions do not place their principal dependence on them, yet they look on them as a very considerable addition to their strength.

But the complete Roman legion, in its own peculiar cohorts, contains within itself the heavy-armed foot, that is: the principes, hastati, triarii, and antefignani, the lightarmed foot, consisting of the ferentarii, archers, slingers, and balistarii, together with the legionary cavalry incorporated with it. These bodies, all actuated with the same spirit, are united inseparably in their various dispositions for forming, encamping and fighting. Thus the legion is compact and perfect in all its parts and, without any foreign assistance, has always been superior to any force that could be brought against it. The Roman greatness is a proof of the excellence of their legions, for with them they always defeated whatever numbers of the enemy they thought fit, or their circumstances gave them an opportunity to engage.

Causes of Decay of the Legion

The name of the legion remains indeed to this day in our armies, but its strength and substance are gone, since by the neglect of our predecessors, honors and preferments, which were formerly the recompenses of merit and long services, were to be attained only by interest and favor. Care is no longer taken to replace the soldiers, who after serving their full time, have received their discharges. The vacancies continually happening by sickness, discharges, desertion and various other casualties, if not supplied every year or even every month, must in time disable the most numerous army. Another cause of the weakness of our legions is that in them the soldiers find the duty hard, the arms heavy, the rewards distant and the discipline severe. To avoid these inconveniences, the young men enlist in the auxiliaries, where the service is less laborious and they have reason to expect more speedy recompenses.

Cato the Elder, who was often Consul and always victorious at the head of the armies, believed he should do his country more essential service by writing on military affairs, than by all his exploits in the field. For the consequences of brave actions are only temporary, while whatever is committed to writing for public good is of lasting benefit. Several others have followed his example, particularly Frontinus, whose elaborate works on this subject were so well received by the Emperor Trajan. These are the authors whose maxims and institutions I have undertaken to abridge in the most faithful and concise manner.

The expense of keeping up good or bad troops is the same; but it depends wholly on You, most August Emperor, to recover the excellent discipline of the ancients and to correct the abuses of later times. This is a reformation the advantages of which will be equally felt by ourselves and our posterity.

The Organization of the Legion

All our writers agree that never more than two legions, besides auxiliaries, were sent under the command of each consul against the most numerous armies of the enemies. Such was the dependence on their discipline and resolution that this number was thought sufficient for any war they were engaged in. I shall therefore explain the organization of the ancient legion according to the military constitution. But if the description appear obscure or imperfect, it is not to be imputed to me, but to the difficulty of the subject itself, which is therefore to be examined with the greater attention. A prince, skilled himself in military affairs, has it in his power to make himself invincible by keeping up whatever number of well disciplined forces he thinks proper.

The recruits having thus been carefully chosen with proper attention to their persons and dispositions, and having been daily exercised for the space of four months at least, the legion is formed by the command and under the auspices of the Emperor. The military mark, which is indelible, is first imprinted on the hands of the new levies, and as their names are inserted in the roll of the legions they take the usual oath, called the military oath. They swear by God, by Christ and by the Holy Ghost; and by the Majesty of the Emperor who, after God, should be the chief object of the love and veneration of mankind. For when he has once received the title of August, his subjects are bound to pay him the most sincere devotion and homage, as the representative of God on earth. And every man, whether in a private or military station, serves God in serving him faithfully who reigns by His authority. The soldiers, therefore, swear they will obey the Emperor willingly and implicitly in all his commands, that they will never desert and will always be ready to sacrifice their lives for the Roman Empire.

The legion should consist of ten cohorts, the first of which exceeds the others both in number and quality of its soldiers, who are selected to serve in it as men of some family and education. This cohort has the care of the eagle, the chief ensign in the Roman armies and the standard of the whole legion, as well as of the images of the emperors which are always considered as sacred. It consists of eleven hundred and five foot and one hundred and thirty-two horse cuirassiers, and is distinguished by the name of the Millarian Cohort. It is the head of the legion and is always first formed on the right of the first line when the legion draws up in order of battle.

The second cohort contains five hundred and fifty-five foot and sixty-six horse, and is called the Quingentarian Cohort. The third is composed of five hundred and fifty-five foot and sixty-six horse, generally chosen men, on account of its situation in the center of the first line. The fourth consists of the same number of five hundred and fifty-five foot and sixty-six horse. The fifth has likewise five hundred and fifty-five foot and sixty-six horse, which should be some of the best men, being posted on the left flank as the first cohort is on the right. These five cohorts compose the first line.

The sixth includes five hundred and fifty-five foot and sixty-six horse, which should be the flower of the young soldiers as it draws up in the rear of the eagle and the images of the emperors, and on the right of the second line. The seventh contains five hundred and fifty-five foot and sixty-six horse. The eighth is composed of five hundred and fifty-five foot and sixty-six horse, all selected troops, as it occupies the center of the second line. The ninth has five hundred and fifty-five foot and sixty-six horse. The tenth consists of the same number of five hundred and fifty-five foot and sixty-six horse and requires good men, as it closes the left flank of the second line. These ten cohorts form the complete legions, consisting in the whole of six thousand one hundred foot and seven hundred and twenty-six horses. A legion should never be composed of a less number of men, but it is sometimes stronger by the addition of other Millarian Cohorts.

The Officers of the Legion

Having shown the ancient establishment of the legion, we shall now explain the names of the principal soldiers or, to use the proper term, the officers, and their ranks according to the present rolls of the legions. The first tribune is appointed by the express commission and choice of the Emperor. The second tribune rises to that rank by length of service. The tribunes are so called from their command over the soldiers, who were at first levied by Romulus out of the different tribes. The officers who in action commanded the orders or divisions are called Ordinarii. The Augustales were added by Augustus to the Ordinarii; and the Flaviales were appointed by Flavius Vespasian to double the number of the Augustales. The eagle-bearers and the image-bearers are those who carry the eagles and images of the Emperors. The Optiones are subaltern officers, so denominated from their being selected by the option of their superior officers, to do their duty as their substitutes or lieutenants in case of sickness or other accident. The ensign-bearers carry the ensigns and are called Draconarii. The Tesserarii deliver the parole and the orders of the general to the different messes of the soldiers. The Campignei or Antefignani are those whose duty it is to keep the proper exercises and discipline among the troops. The Metatores are ordered before the army to fix on the ground for its encampments. The Beneficiarii are so named from their owing their promotion to the benefit or interest of the Tribunes. The Librarii keep the legionary accounts. The Tubicines, Cornicines, and Buccinatores derive their appellations from blowing the trumpet, cornet, and buccina. Those who, expert in their exercises, receive a double allowance of provisions, are called Armaturae Duplares, and those who have but a single portion, Simplares. The Mensores mark out the ground by measure for the tents in an encampment, and assign the troops their respective quarters in garrison. The Torquati, so denominated from the gold collars given them in reward for their bravery, had besides this honor different allowances. Those who received double were called Torquati Duplares, and those who had only single, Simplares. There were, for the same reason, Candidatii Duplares, and Candidatii Simplares. These are the principal soldiers or officers distinguished by their rank and privileges thereto annexed. The rest are called Munifices, or working soldiers, from their being obliged to every kind of military work without exception. Formerly it was the rule that the first Princeps of the legion should be promoted regularly to the rank of Centurion of the Primiple. He not only was entrusted with the eagle but commanded four centuries, that is, four hundred men in the first line. As head of the legion he had appointments of great honor and profit. The first Hastatus had the command of two centuries or two hundred men in the second line, and is now called Ducenarius. The Princeps of the first cohort commanded a century and a half, that is, one hundred and fifty men, and kept in a great measure the general detail of the legion. The second Hastatus had likewise a century and a half, or one hundred and fifty men. The first Triarius had the command of one hundred men. Thus the ten centuries of the first cohort were commanded by five Ordinarii, who by the ancient establishment enjoyed great honors and emoluments that were annexed to this rank in order to inspire the soldiers of the legions with emulation to attain such ample and considerable rewards. They had also Centurions appointed to each century, now called Centenarii and Decani, who commanded ten men, now called heads of messes. The second cohort had five Centurions; and all the rest to the tenth inclusively the same number. In the whole legion there were fifty-five.

Lieutenants of consular dignity were formerly sent to command in the armies under the general, and their authority extended over both the legions and auxiliaries in peace and war. Instead of these officers, persons of high rank are now substituted with the title of Masters of the Forces. They are not limited to the command of two legions only, but have often a greater number. But the peculiar officer of the legion was the Praefect, who was always a count of the first order. On him the chief command devolved in the absence of the lieutenant. The Tribunes, Centurions, and all the soldiers in general were under his orders: He gave out the parole and order for the march and for the guards. And if a soldier committed a crime, by his authority the Tribune adjudged him to punishment. He had charge of the arms, horses, clothing and provisions. It was also his duty to keep both the legionary horse and foot in daily exercise and to maintain the strictest discipline. He ought to be a careful and diligent officer, as the sole charge of forming the legion to regularity and obedience depended on him and the excellence of the soldiers redounded entirely to his own honor and credit.

The Praefect of the camp, though inferior in rank to the former, had a post of no small importance. The position of the camp, the direction of the entrenchments, the inspection of the tents or huts of the soldiers and the baggage were comprehended in his province. His authority extended over the sick, and the physicians who had the care of them; and he regulated the expenses relative thereto. He had the charge of providing carriages, bathorses and the proper tools for sawing and cutting wood, digging trenches, raising parapets, sinking wells and bringing water into the camp. He likewise had the care of furnishing the troops with wood and straw, as well as the rams, onagri, balistae and all the other engines of war under his direction. This post was always conferred on an officer of great skill, experience and long service, and who consequently was capable of instructing others in those branches of the profession in which he had distinguished himself.

The Praefect of the Workmen

The legion had a train of joiners, masons, carpenters, smiths, painters, and workmen of every kind for the construction of barracks in the winter-camps and for making or repairing the wooden towers, arms, carriages and the various sorts of machines and engines for the attack or defense of places. They had also traveling workshops in which they made shields, cuirasses, helmets, bows, arrows, javelins and offensive and defensive arms of all kinds. The ancients made it their chief care to have every thing for the service of the army within the camp. They even had a body of miners who, by working under ground and piercing the foundations of walls, according to the practice of the Beffi, penetrated into the body of a place. All these were under the direction of the officer called the praefect of the workmen.

The Tribune of the Soldiers

We have observed that the legions had ten cohorts, the first of which, called the Millarian Cohort, was composed of men selected on account of their circumstances, birth, education, person and bravery. The tribune who commanded them was likewise distinguished for his skill in his exercises, for the advantages of his person and the integrity of his manners. The other cohorts were commanded, according to the Emperor's pleasure, either by tribunes or other officers commissioned for that purpose. In former times the discipline was so strict that the tribunes or officers abovementioned not only caused the troops under their command to be exercised daily in their presence, but were themselves so perfect in their military exercises as to set them the example. Nothing does so much honor to the abilities or application of the tribune as the appearance and discipline of the soldiers, when their apparel is neat and clean, their arms bright and in good order and when they perform their exercises and evolutions with dexterity.

Centuries and Ensigns of the Foot

The chief ensign of the whole legion is the eagle and is carried by the eagle-bearer. Each cohort has also its own peculiar ensign, the Dragon, carried by the Draconarius. The ancients, knowing the ranks were easily disordered in the confusion of action, divided the cohorts into centuries and gave each century an ensign inscribed with the number both of the cohort and century so that the men keeping it in sight might be prevented from separating from their comrades in the greatest tumults. Besides the centurions, now called centenarii, were distinguished by different crests on their helmets, to be more easily known by the soldiers of their respective centuries. These precautions prevented any mistake, as every century was guided not only by its own ensign but likewise by the peculiar form of the helmet of its commanding officers. The centuries were also subdivided into messes of ten men each who lay in the same tent and were under orders and inspection of a Decanus or head of the mess. These messes were also called Maniples from their constant custom of fighting together in the same company or division.

Legionary Troops of Horse

As the divisions of the infantry are called centuries, so those of the cavalry are called troops. A troop consists of thirty-two men and is commanded by a Decurion. Every century has its ensign and every troop its Standard. The centurion in the infantry is chosen for his size, strength and deXterity in throwing his missile weapons and for his skill in the use of his sword and shield; in short for his expertness in all the exercises. He is to be vigilant, temperate, accive and readier to execute the orders he receives than to talk; Strict in exercising and keeping up proper discipline among his soldiers, in obliging them to appear clean and well-dressed and to have their arms constantly rubbed and bright. In like manner the Decurion is to be preferred to the command of a troop for his activity and address in mounting his horse completely armed; for his skill in riding and in the use of the lance and bow; for his attencion in forming his men to all the evolutions of the cavalry; and for his care in obliging them to keep their cuirasses, lances and helmets always bright and in good order. The splendor of the arms has no inconsiderable effect in striking terror into an enemy. Can that man be reckoned a good soldier who through negligence suffers his arms to be spoiled by dirt and rust? In short, it is the duty of the Decurion to be attentive to whatever concerns the health or discipline of the men or horses in his troop.

Drawing Up a Legion in Order of Battle

We shall exemplify the manner of drawing up an army in order of battle in the instance of one legion, which may serve for any number. The cavalry are posted on the wings. The infantry begin to form on a line with the :first cohort on the right. The second cohort draws up on the left of the first; the third occupies the center; the fourth is posted next; and the fifth closes the left flank. The ordinarii, the other officers and the soldiers of the first line, ranged before and round the ensigns, were called the principes. They were all heavy armed troops and had helmets, cuirasses, greaves, and shields. Their offensive weapons were large swords, called spathae, and smaller ones called semispathae together with five loaded javelins in the concavity of the shield, which they threw at the first charge. They had likewise two other javelins, the largest of which was composed of a staff five feet and a half long and a triangular head of iron nine inches long. This was formerly called the pilum, but now it is known by the name of spiculum. The soldiers were particularly exercised in the use of this weapon, because when thrown with force and skill it often penetrated the shields of the foot and the cuirasses of the horse. The other javelin was of smaller size; its triangular point was only five inches long and the staff three feet and one half. It was anciently called verriculum but now verutum.

The first line, as I said before, was composed of the principes; the hastati formed the second and were armed in the same manner. In the second line the sixth cohort was posted on the right flank, with the seventh on its left; the eighth drew up in the center; the ninth was the next; and the tenth always closed the left flank. In the rear of these two lines were the ferentarii, light infantry and the troops armed with shields, loaded javelins, swords and common missile weapons, much in the same manner as our modern soldiers. This was also the post of the archers who had helmets, cuirasses, swords, bows and arrows; of the slingers who threw stones with the common sling or with the fustibalus; and of the tragularii who annoyed the enemy with arrows from the manubalistae or arcubalistae.

In the rear of all the lines, the triarii, completely armed, were drawn up. They had shields, cuirasses, helmets, greaves, swords, daggers, loaded javelins, and two of the common missile weapons. They rested during the acnon on one knee, so that if the first lines were obliged to give way, they might be fresh when brought up to the charge, and thereby retrieve what was lost and recover the victory. All the ensigns though, of the infantry, wore cuirasses of a smaller sort and covered their helmets with the shaggy skins of beasts to make themselves appear more terrible to the enemy. But the centurions had complete cuirasses, shields, and helmets of iron, the crest of which, placed transversely thereon, were ornamented with silver that they might be more easily distinguished by their respective soldiers.

The following disposition deserves the greatest attention. In the beginning of an engagement, the first and second lines remained immovable on their ground, and the trairii in their usual positions. The light-armed troops, composed as above mentioned, advanced in the front of the line, and attacked the enemy. If they could make them give way, they pursued them; but if they were repulsed by superior bravery or numbers, they retired behind their own heavy armed infantry, which appeared like a wall of iron and renewed the action, at first with their missile weapons, then sword in hand. If they broke the enemy they never pursued them, least they should break their ranks or throw the line into confusion, and lest the enemy, taking advantage of their disorder, should return to the attack and destroy them without difficulty. The pursuit therefore was entirely left to the light-armed troops and the cavalry. By these precautions and dispositions the legion was victorious without danger, or if the contrary happened, was preserved without any considerable loss, for as it is not calculated for pursuit, it is likewise not easily thrown into disorder.

Names of Soldiers Inscribed on Their Shields

Lest the soldiers in the confusion of battle should be separated from their comrades, every cohort had its shields painted in a manner peculiar to itself. The name of each soldier was also written on his shield, together with the number of the cohort and century to which he belonged. From this description we may compare the legion, when in proper order, to a well fortified city as containing within itself every thing requisite in war, wherever it moved. It was secure from any sudden attempt or surprise of an enemy by its expeditious method of entrenching its camp even in the open plains and it was always provided with troops and arms of every kind. To be victorious, therefore, over our enemies in the field, we must unanimously supplicate heaven to dispose the Emperor to reform the abuses in raising our levies and to recruit our legions after the method of the ancients. The same care in choosing and instructing our young soldiers in all military exercises and drills will soon make them equal to the old Roman troops who subdued the whole world. Nor let this alteration and loss of ancient discipline in any way affect Your Majesty, since it is a happiness reserved for You alone both to restore the ancient ordinances and establish new ones for the public welfare. Every work before the attempt carries in it an appearance of difficulty; but in this case, if the levies are made by careful and experienced officers, an army may be raised, disciplined and rendered fit for service in a very short time; for the necessary expenses once provided, diligence soon effects whatever it undertakes.

Records and Accounts

Several posts in the legion requiring men of some education, the superintendents of the levies should select some recruits for their skill in writing and accounts, besides the qualification to be attended to in general, such as size, strength and proper disposition for the service. For the whole detail of the legion, including the lists of the soldiers exempted from duty on private accounts, the rosters for their tour of military duties and their pay lists, is daily entered in the legionary books and kept we may almost say, with greater exactness than the regulations of provisions or other civil matters in the registers of the police. The daily guards in time of peace, the advanced guards and outposts in time of war, which are mounted regularly by the centuries and messes in their turns, are likewise punctually kept in rolls for that purpose, with the name of each soldier whose tour is past, that no one may have injustice done him or be excused from his duty by favor.

They are also exact in entering the time and limitation of furloughs, which formerly were never granted without difficulty and only on real and urgent business. They then never suffered the soldiers to attend on any private person or to concern themselves in private occupations, thinking it absurd and improper that the Emperor's soldiers, clothed and subsisted at the public expense, should follow any other profession. Some soldiers, however, were allowed for the service of the praefects, tribunes and even of the other officers, our of the number of the accensi or such as were raised after the legion was complete. These latter are now called supernumeraries. The regular troops were obliged to carry their wood, hay, water and straw into the camp themselves. From such kind of services they were called munifices.

Soldier's Deposits

The institution of the ancients which obliged the soldiers to deposit half of every donative they received at the colors was wise and judicious; the intent was to preserve it for their use so that they might not squander it in extravagance or idle expense. For most men, particularly the poorer sort, soon spend whatever they can get. A reserve of this kind therefore is evidently of the greatest service to the soldiers themselves; since they are maintained at the public expense, their military stock by this method is continually increasing. The soldier who knows all his fortune is deposited at his colors, entertains no thoughts of desertion, conceives a greater affection for them and fights with greater intrepidity in their defense. He is also prompted thereto by interest, the most prevailing consideration among men. This money was contained in ten bags, one for each cohort. There was an eleventh bag also for a small contribution from the whole legion, as a common fund to defray the expense of burial of any of their deceased comrades. These collections were kept in baskets in the custody of the ensigns, chosen for their integrity and capacity, and answerable for the trust and obliged to account with every man for his own proportion.

Promotion in the Legion

Heaven certainly inspired the Romans with the organization of the legion, so superior does it seem to human invention. Such is the arrangement and disposition of the ten cohorts that compose it, as to appear one perfect body and form one complete whole. A soldier, as he advances in rank, proceeds as it were by rotation through the different degrees of the several cohorts in such a manner that one who is promoted passes from the first cohort to the tenth, and returns again regularly through all the others with a continual increase of rank and pay to the first. Thus the centurion of the primiple, after having commanded in the different ranks of every cohort, attains that great dignity in the first with infinite advantages from the whole legion. The chief praefect of the Praetorian Guards rises by the same method of rotation to that lucrative and honorable rank. Thus the legionary horse contract an affection for the foot of their own cohorts, notwithstanding the natural antipathy existing between the two corps. And this connection establishes a reciprocal attachment and union between all the cohorts and the cavalry and infantry of the legion.

Legionary Music

The music of the legion consists of trumpets, cornets and buccinae. The trumpet sounds the charge and the retreat. The cornets are used only to regulate the motions of the colors; the trumpets serve when the soldiers are ordered out to any work without the colors; but in time of action, the trumpets and cornets sound together. The classicum, which is a particular sound of the buccina or horn, is appropriated to the commander-in-chief and is used in the presence of the general, or at the execution of a soldier, as a mark of its being done by his authority. The ordinary guards and outposts are always mounted and relieved by the sound of trumpet, which also directs the motions of the soldiers on working parties and on field days. The cornets sound whenever the colors are to be struck or planted. These rules must be punctually observed in all exercises and reviews so that the soldiers may be ready to obey them in action without hesitation according to the general's orders either to charge or halt, to pursue the enemy or to retire. F or reason will convince us that what is necessary to be performed in the heat of action should constantly be practiced in the leisure of peace.

The Drilling of the Troops

The organization of the legion being thus explained, let us return to the drills. The younger soldiers and recruits went through their drills of every kind every morning and afternoon and the veterans and most expert regularly once a day. Length of service or age alone will never form a military man, for after serving many years an undisciplined soldier is still a novice in his profession. Not only those under the masters at arms, but all the soldiers in general, were formerly trained incessantly in those drills which now are only exhibited as shows in the circus for particular solemnities. By practice only can be acquired agility of body and the skill requisite to engage an enemy with advantage, especially in close fight. But the most essential point of all is to teach soldiers to keep their ranks and never abandon their colors in the most difficult evolutions. Men thus trained are never at a loss amidst the greatest confusion of numbers.

The recruits likewise are to be exercised with wooden swords at the post, to be taught to attack this imaginary antagonist on all sides and to aim at the sides, feet or head, both with the point and edge of the sword. They must be instructed how to spring forward to give the blow, to rise with a bound above the shield and then to sink down and shelter themselves under cover of it, and how to advance and retire. They must also throw their javelins at the post from a considerable distance in order to acquire a good aim and strengthen the arm.

The archers and slingers set up bundles of twigs or straw for marks, and generally strike them with arrows and with stones from the fustiablus at the distance of six hundred feet. They acquired coolness and exactness in acnon from familiar custom and exercise in the field. The slingers should be taught to whirl the sling but once about the head before they cast the stone. Formerly all soldiers were trained to the praccice of throwing stones of a pound weight with the hand, as this was thought a readier method since it did not require a sling. The use of the common missile weapons and loaded javelins was another part of the drill strictly attended to.

To continue this drill without interruption during the winter, they erected for the cavalry porticos or riding halls covered with tiles or shingles, and if they were not to be procured, with reeds, rushes or thatch. Large open halls were likewise constructed in the same manner for the use of the infantry. By these means the troops were provided with places of drill sheltered from bad weather. But even in winter, if it did not rain or snow, they were obliged to perform their drills in the field, lest an intermission of discipline should affect both the courage and constitution of the soldier. In short, both legionary and auxiliary troops should continually be drilled in cutting wood, carrying burdens, passing ditches, swimming in the sea or in rivers, marching in the full step and even running with their arms and baggage, so that, inured to labor in peace, they may find no difficulty in war. For, as the well trained soldier is eager for action, so does the untaught fear it. In war discipline is superior to strength; but if that discipline is neglected, there is no longer any difference between the soldier and the peasant. The old maxim is certain that the very essence of an art consists in constant practice.

Machines and Tools of the Legion

The legion owes its success to its arms and machines, as well as to the number and bravery of its soldiers. In the first place every century has a balista mounted on a carriage drawn by mules and served by a mess, that is by ten men from the century to which it belongs. The larger these engines are, the greater distance they carry and with the greater force. They are used not only to defend the entrenchments of camps, but are also placed in the field in the rear of the heavy armed infantry. And such is the violence with which they throw the darts that neither the cuirasses of the horse nor shields of the foot can resist them. The number of these engines in a legion is fifty-five. Besides these are ten onagri, one for each cohort; they are drawn ready armed on carriages by oxen; in case of an attack, they defend the works of the camp by throwing stones as the balistae do darts.

The legion carries with it a number of small boats, each hollowed out of a single piece of timber, with long cables and sometimes iron chains to fasten them together. These boats, joined and covered with planks, serve as bridges over unfordable rivers, on which both cavalry and infantry pass without danger. The legion is provided with iron hooks, called wolves, and iron scythes fixed to the ends of long poles; and with forks, spades, shovels, pickaxes, wheelbarrows and baskets for digging and transporting earth; together with hatchets, axes and saws for cutting wood. Besides which, a train of workmen attend on it furnished with all instruments necessary for the construction of tortoises, musculi, rams, vines, moving towers and other machines for the attack of places. As the enumeration of all the particulars of this sort would be too tedious, I shall only observe that the legion should carry with it wherever it moves, whatever is necessary for every kind of service so that the encampments may have all the strength and conveniences of a fortified city.

6

DISPOSITIONS FOR ACTION

To the Emperor Valentinian

The Athenians and Lacedaemonians were masters of Greece before the Macedonians, as history informs us. The Athenians excelled not only in war but in other arts and sciences. The Lacedaemonians made war their chief study. They are affirmed to be the first who reasoned on the events of battles and committed their observations thereon to writing with such success as to reduce the military art, before considered as totally dependent on courage or fortune, to certain rules and fixed principles. As a consequence they established schools of tactics for the instruction of youth in all the maneuvers of war. How worthy of admiration are these people for particularly applying themselves to the study of an art, without which no other art can possibly exist. The Romans followed their example, and both practiced their institutions in their armies and preserved them in their writings. These are the maxims and instructions dispersed through the works of different authors, which Your Majesty has ordered me to abridge, since the perusal of the whole would be too tedious, and the authority of only a part unsatisfactory. The effect of the Lacedaemonian skill in dispositions for general actions appears evidently in the single instance of Xantippus, who assisted the Carthaginians after the repeated ruin of their armies. And merely superior skill and conduct defeated Attilius Regulus at the head of a Roman army, till that time always victorious. Xantippus took him prisoner and thus terminated the war by a single action. Hannibal, also, before he set out on his expedition into Italy, chose a Lacedaemonian for his counsellor in military operations; and by his advice, though inferior to the Romans both in number and strength, overthrew so many consuls and such mighty legions. He, therefore, who desires peace, should prepare for war. He who aspires to victory, should spare no pains to form his soldiers. And he who hopes for success, should fight on principle, not chance. No one dares to offend or insult a power of known superiority in action.

The Number Which Should Compose an Army

The first book treats of the choice and exercises of new levies; the second explains the establishment of the legion and the method of discipline; and the third contains the dispositions for action. By this methodical progression, the following instructions on general actions and means of victory will be better understood and of greater use. By an army is meant a number of troops, legions and auxiliaries, cavalry and infantry, assembled to make war. This number is limited by judges of the profession. The defeats of Xerxes, Darius, Mithridates and other monarchs who brought innumerable multitudes into the field, plainly show that the destruction of such prodigious armies is owing more to their own numbers than to the bravery of their enemies. An army too numerous is subject to many dangers and inconveniences. Its bulk makes it slow and unwieldy in its motions; and as it is obliged to march in columns of great length, it is exposed to the risk of being continually harassed and insulted by inconsiderable parties of the enemy. The incumbrance of the baggage is often an occasion of its being surprised in its passage through difficult places or over rivers. The difficulty of providing forage for such numbers of horses and other beasts of burden is very great. Besides, scarcity of provisions, which is to be carefully guarded against in all expeditions, soon ruins such large armies where the consumption is so prodigious, that notwithstanding the greatest care in filling the magazines they must begin to fail in a short time. And sometimes they unavoidably will be distressed for want of water. But, if unfortunately this immense army should be defeated, the numbers lost must necessarily be very great, and the remainder, who save themselves by flight, too much dispirited to be brought again to action.

The ancients, taught by experience, preferred discipline to numbers. In wars of lesser importance they thought one legion with auxiliaries, that is, ten thousand foot and two thousand horse, sufficient. And they often gave the command to a praeter as to a general of the second rank. When the preparations of the enemy were formidable, they sent a general of consular dignity with twenty thousand foot and four thousand horse. In our times this command was given to a count of the first order. But when there happened any dangerous insurrection supported by infinite multitudes of fierce and barbarous nations, on such emergencies they took the field with two armies under two consuls, who were charged, both singly and jointly, to take care to preserve the Republic from danger. In short, by this management, the Romans, almost continually engaged in war with different nations in different parts of the world, found themselves able to oppose them in every quarter. The excellence of their discipline made their small armies sufficient to encounter all their enemies with success. But it was an invariable rule in their armies that the number of allies or auxiliaries should never exceed that of the Roman citizens.

Means of Preserving It in Health

The next article is of the greatest importance: the means of preserving the health of the troops. This depends on the choice of situation and water, on the season of the year, medicine, and exercise. As to the situation, the army should never continue in the neighborhood of unwholesome marshes any length of time, or on dry plains or eminences without some sort of shade or shelter. In the summer, the troops should never encamp without tents. And their marches, in that season of the year when the heat is excessive, should begin by break of day so that they may arrive at the place of destination in good time. Otherwise they will contract diseases from the heat of the weather and the fatigue of the march. In severe winter they should never march in the night in frost and snow, or be exposed to want of wood or clothes. A soldier, starved with cold, can neither be healthy nor fit for service. The water must be wholesome and not marshy. Bad water is a kind of poison and the cause of epidemic distempers.

It is the duty of the officers of the legion, of the tribunes, and even of the commander-in-chief himself, to take care that the sick soldiers are supplied with proper diet and diligently attended by the physicians. For little can be expected from men who have both the enemy and diseases to struggle with. However, the best judges of the service have always been of the opinion that daily practice of the military exercises is much more efficacious toward the health of an army than all the art of medicine.

For this reason they exercised their infantry without intermission. If it rained or snowed, they performed under cover; and i11 fine weather, in the field. They also were assiduous in exercising their cavalry, not only in plains, but also on uneven ground, broken and cut with ditches. The horses as well as the men were thus trained, both on the above mentioned account and to prepare them for action. Hence we may perceive the importance and necessity of a strict observance of the military exercises in an army, since health in the camp and victory in the field depend on them. If a numerous army continues long in one place in the summer or in the autumn, the waters become corrupt and the air infected. Malignant and fatal distempers proceed from this and can be avoided only by frequent changes of encampments.

Care to Provide Forage and Provisions

Famine makes greater havoc in an army than the enemy, and is more terrible than the sword. Time and opportunity may help to retrieve other misfortunes, but where forage and provisions have not been carefully provided, the evil is without remedy. The main and principal point in war is to secure plenty of provisions and to destroy the enemy by famine. An exact calculation must therefore be made before the commencement of the war as to the number of troops and the expenses incident thereto, so that the provinces may in plenty of time furnish the forage, corn, and all other kinds of provisions demanded of them to be transported. They must be in more than sufficient quantity, and gathered into the strongest and most convenient cities before the opening of the campaign. If the provinces cannot raise their quotas in kind, they must commute for them in money to be employed in procuring all things requisite for the service. For the possessions of the subjects cannot be kept secure otherwise than by the defense of arms.

These precautions often become doubly necessary as a siege is sometimes protracted beyond expectation, the besiegers resolving to suffer themselves all the inconveniences of want sooner than raise the siege, if they have any hopes of reducing the place by famine. Edicts should be issued out requiring the country people to convey their cattle, grain, wine and all kinds of provisions that may be of service to the enemy, into garrisoned fortresses or into the safest cities. And if they do not comply with the order, proper officers are to appointed to compel them to do it. The inhabitants of the province must likewise be obliged to retire with their effects into some fortified place before the irruption of the enemy. The fortifications and all the machines of different kinds must also be examined and repaired in time. For if you are once surprised by the enemy before you are in a proper posture of defense, you are thrown into irrecoverable confusion, and you can no longer draw any assistance from the neighboring places, all communication with them being cut off. But a faithful management of the magazines and a frugal distribution of the provisions, with proper precautions taken at first, will insure sufficient plenty. When provisions once begin to fail, parsimony is ill-timed and comes too late.

On difficult expeditions the ancients distributed the provisions at a fixed allowance to each man without distinction of rank; and when the emergency was past, the government accounted for the full proportions. The troops should never want wood and forage in winter or water in summer. They should have corn, wine, vinegar, and even salt, in plenty at all times. Cities and fortresses are garrisoned by such mcn as are least fit for thc service of the field. They arc provided with all sorts of arnms, arrows, fustibali, slings, stones, onagri and balistae for their defense. Great caution is requisite that the unsuspecting simplicity of the inhabitants be not imposed on by the treachery or perjury of the enemy, for pretended conferences and deceitful appearance of truces have often been more fatal than force. By observing the foregoing precautions, the besieged may have it in their power to ruin the enemy by famine, if he keeps his troops together, and if he divides them, by frequent sallies and surprises.

Methods to Prevent Mutiny in an Army

An army drawn together from different parts sometimes is disposed to mutiny. And the troops, though not inclined to fight, pretend to be angry at not being led against the enemy. Such seditious dispositions principally show themselves in those who have lived in their quarters in idleness and effeminacy. These men, unaccustomed to the necessary fatigue of the field, are disgusted at its severity. Their ignorance of discipline makes them afraid of action and inspires them with insolence.

There are several remedies for this evil. While the troops are yet separated and each corps continues in its respective quarters, let the tribunes, their lieutenants and the officers in genera4 make it their business to keep up so strict a discipline as to leave them no room to harbor any thoughts but of submission and obedience. Let them be constantly employed either in field days or in the inspection of their arms. They should not be allowed to be absent on furlough. They should be frequently called by roll and trained to be exact in the observance of every signal. Let them be exercised in the use of the bow, in throwing missile weapons and stones, both with the hand and sling, and with the wooden sword at the post; let all this be continually repeated and let them be often kept under arms till they are tired. Let them be exercised in running and leaping to facilitate the passing of ditches. And if their quarters are near the sea or a river, let them all, without exception, be obliged in the summer to have the frequent practice of swimming. Let them be accustomed to march through thickets, inclosures and broken grounds, to fell trees and cut out timber, to break ground and to defend a post against their comrades who are to endeavor to dispossess them; and in the encounter each party should use their shields to dislodge and bear down their antagonists. All the different kinds of troops thus trained and exercised in their quarters will find themselves inspired with emulation for glory and eagerness for action when they come to take the field. In short, a soldier who has proper confidence in his own skill and strength, entertains no thought of mutiny.

A general should be attentive to discover the turbulent and seditious soldiers in the army, legions or auxiliaries, cavalry or infantry. He should endeavor to procure his intelligence not from informers, but from the tribunes, their lieutenants and other officers of undoubted veracity. It would then be prudent in him to separate them from the rest under pretence of some service agreeable to them, or detach them to garrison cities or castles, but with such address that though he wants to get rid of them, they may think themselves employed by preference and favor. A multitude never broke out into open sedition at once and with unanimous consent. They are prepared and excited by some few mutineers, who hope to secure impunity for their crimes by the number of their associates. But if the height of the mutiny requires violent remedies, it will be most advisable, after the manner of the ancients, to punish the ring-leaders only in order that, though few suffer, all may be terrified by the example. But it is much more to the credit of a general to form his troops to submission and obedience by habit and discipline than to be obliged to force them to their duty by the terror of punishment.

Marches in the Neighborhood of the Enemy

It is asserted by those who have made the profession their study that an army is exposed to more danger on marches than in battles. In an engagement the men are properly armed, they see their enemies before them and come prepared to fight. But on a march the soldier is less on his guard, has not his arms always ready and is thrown into disorder by a sudden attack or ambuscade. A general, therefore, cannot be too careful and diligent in taking necessary precautions to prevent a surprise on the march and in making proper dispositions to repulse the enemy, in case of such accident, without loss.

In the first place, he should have an exact description of the country that is. the seat of war, in which the distances of places specified by the number of miles, the nature of the roads, the shortest routes, by-roads, mountains and rivers, should be correctly inserted. We are told that the greatest generals have carried their precautions on this head so far that, not satisfied with the simple description of the country wherein they were engaged, they caused plans to be taken of it on the spot, that they might regulate their marches by the eye with greater safety. A general should also inform himself of all these particulars from persons of sense and reputation well acquainted with the country by examining them separately at first, and then comparing their accounts in order to come at the truth with certainty.

If any difficulty arises about the choice of roads, he should procure proper and skillful guides. He should put them under a guard and spare neither promises nor threat to induce them to be faithful. They will acquit themselves well when they know it is impossible to escape and are certain of being rewarded for their fidelity or punished for their perfidy. He must be sure of their capacity and experience, that the whole army be not brought into danger by the errors of two or three persons. For sometimes the common sort of people imagine they know what they really do not, and through ignorance promise more than they can perform.

But of all precautions the most important is to keep entirely secret which way or by what route the army is to march. For the security of an expedition depends on the concealment of all motions from the enemy. The figure of the Minotaur was anciently among the legionary ensigns, signifying that this monster, according to the fable, was concealed in the most secret recesses and windings of the labyrinth, just as the designs of a general should always be impenetrable. When the enemy has no intimation of a march, it is made with security; but as sometimes the scouts either suspect or discover the decampment, or traitors or deserters give intelligence thereof, it will be proper to mention the method of acting in case of an attack on the march.

The general, before he puts his troops in motion, should send out detachments of trusty and experienced soldiers well mounted, to reconnoiter the places through which he is to march, in front, in rear, and on the right and left, lest he should fall into ambuscades. The night is safer and more advantageous for your spies to do their business in than day, for if they are taken prisoners, you have, as it were, betrayed yourself. After this, the cavalry should march off first, then the infantry; the baggage, bat horses, servants and carriages follow in the center; and part of the best cavalry and infantry come in the rear, since it is oftener attacked on a march than the front. The flanks of the baggage, exposed to frequent ambuscades, must also be covered with a sufficient guard to secure them. But above all, the part where the enemy is most expected must be reinforced with some of the best cavalry, light infantry and foot archers.

If surrounded on all sides by the enemy, you must

make dispositions to receive them wherever they come, and the soldiers should be cautioned beforehand to keep their arms in their hands, and to be ready in order to prevent the bad effects of a sudden attack. Men are frightened and thrown into disorder by sudden accidents and surprises of no consequence when foreseen. The ancients were very careful that the servants or followers of the army, if wounded or frightened by the noise of the action, might not disorder the troops while engaged, and also to prevent their either straggling or crowding one another too much, which might incommode their own men and give advantage to the enemy. They ranged the baggage, therefore, in the same manner as the regular troops under particular ensigns. They selected from among the servants the most proper and experienced and gave them the command of a number of servants and boys, not exceeding two hundred, and their ensigns directed them where to assemble the baggage. Proper intervals should always be kept between the baggage and the troops, that the latter may not be embarrassed for want of room in case of an attack during the march. The manner and disposition of defense must be varied according to the difference of ground. In an open country you are more liable to be attacked by horse than foot. But in a woody, mountainous or marshy situation, the danger to be apprehended is from foot. Some of the divisions being apt through negligence to move too fast, and others too slow, great care is to be taken to prevent the army from being broken or from running into too great a length, as the enemy would instantly take advantage of the neglect and penetrate without difficulty.

The tribunes, their lieutenants or the masters at

arms of most experience, must therefore be posted at proper distances, in order to halt those who advance too fast and quicken such as move too slow. The men at too great a distance in the front, on the appearance of an enemy, are more disposed to fly than to join their comrades. And those too far behind, destitute of assistance, fall a sacrifice to the enemy and their own despair. The enemy, it may be concluded, will either plant ambuscades or make his attack by open force, according to the advantage of the ground. Circumspection in examining every place will be a security against concealed danger; and an ambuscade, if discovered and promptly surrounded, will return the intended mischief with interest.

If the enemy prepare to fall upon you by open force in a mountainous country, detachments must be sent forward to occupy the highest eminences, so that on their arrival they may not dare to attack you under such a disadvantage of ground, your troops being posted so much above their and presenting a front ready for their reception. It is better to send men forward with hatchets and other tools in order to open ways that are narrow but safe, without regard to the labor, rather than to run any risk in the finest roads. It is necessary to be well acquainted whether the enemy usually make their attempts in the night, at break of day or in the hours of refreshment or rest; and by knowledge of their customs to guard against what we find their general practice. We must also inform ourselves whether they are strongest in infantry or cavalry; whether their cavalry is chiefly armed with lances or with bows; and whether their principal strength consists in their numbers or the excellence of their arms. All of this will enable us to take the most proper measures to distress them and for our advantage. When we have a design in view, we must consider whether it will be most advisable to begin the march by day or by night; we must calculate the distance of the places we want to reach; and take such precautions that in summer the troops may not suffer for want of water on their march, nor be obstructed in winter by impassable morasses or torrents, as these would expose the army to great danger before it could arrive at the place of its destination. As it highly concerns us to guard against these inconveniences with prudence, so it would be inexcusible not to take advantage of an enemy that fell into them through ignorance or negligence. Our spies should be constantly abroad; we should spare no pains in tampering with their men, and give all manner of encouragement to deserters. By these means we may get intelligence of their present or future designs. And we should constantly keep in readiness some detachments of cavalry and light infantry, to fall upon them when they least expect it, either on the march, or when foraging or marauding.

Passages of Rivers

The passages of rivers are very dangerous without great precaution. In crossing broad or rapid streams, the baggage, servants, and sometimes the most indolent soldiers are in danger of being lost. Having first sounded the ford, two lines of the best mounted cavalry are ranged at a convenient distance entirely across the river, so that the infantry and baggage may pass between them. The line above the ford breaks the violence of the stream, and the line below recovers and transports the men carried away by the current. When the river is too deep to be forded either by the cavalry or infantry, the water is drawn off, if it runs in a plain, by cutting a great number of trenches, and thus it is passed with ease.

Navigable rivers are passed by means of piles driven into the bottom and floored with planks; or in a sudden emergency by fastening together a number of empty casks and covering them with boards. The cavalry, throwing off their accoutrements, make small floats of dry reeds or rushes on which they lay their rams and cuirasses to preserve them from being wet. They themselves swim their horses across the river and draw the floats after them by a leather thong.

But the most commodious invention is that of the small boats hollowed out of one piece of timber and very light both by their make and the quality of the wood. The army always has a number of these boats upon carriages, together with a sufficient quantity of planks and iron nails. Thus with the help of cables to lash the boats together, a bridge is instantly constructed, which for the time has the solidity of a bridge of stone.

As the enemy generally endeavor to fall upon an army at the passage of a river either by surprise or ambuscade, it is necessary to secure both sides thereof by strong detachments so that the troops may not be attacked and defeated while separated by the channel of the river. But it is still safer to palisade both the posts, since this will enable you to sustain any attempt without much loss. If the bridge is wanted, not only for the present transportation of the troops but also for their return and for convoys, it will be proper to throw up works with large ditches to cover each head of the bridge, with a sufficient number of men to defend them as long as the circumstances of affairs require.

Rules for Encamping an Army

An army on the march cannot expect always to find walled cities for quarters, and it is very imprudent and dangerous to encamp in a straggling manner without some sort of entrenchment. It is an easy matter to surprise troops while refreshing themselves or dispersed in the different occupations of the service. The darkness of night, the necessity of sleep and the dispersion of the horses at pasture afford opportunities of surprise. A good situation for a camp is not sufficient; we must choose the very best that can be found lest, having failed to occupy a more advantageous post the enemy should get possession of it to our great detriment.

An army should not encamp in summer near bad waters or far from good ones, nor in winter in a situation without plenty of forage and wood. The camp should not be liable to sudden inundations. The avenues should not be too steep and narrow lest, if invested, the troops should find it difficult to make their retreat; nor should it be commanded by any eminences from which it may be annoyed by the enemy's weapons. After these precautions, the camp is formed square, round, triangular or oblong, according to the nature of the ground. For the form of a camp does not constitute its goodness. Those camps, however, are thought best where the length is one third more than the depth. The dimensions must be exactly computed by the engineers, so that the size of the camp may be proportioned to the number of troops. A camp which is too confined will not permit the troops to perform their movements with freedom, and one which is too extensive divides them too much. There are three methods of entrenching a camp. The first is for the case when the army is on the march and will continue in the camp for only one night. They then throw up a slight parapet of turf and plant it with a row of palisades or caltrops* of wood. The sods are cut with iron instruments. If the earth is held strongly together by the roots of the grass, they are cut in the form of a brick a foot and one half high, a foot broad and a foot and one half long. If the earth is so loose that the turf cannot be cut in this form, they run a slight trench round the camp, five feet broad and three feet deep. The earth taken from the trench forms a parapet on the inside and this secures the army from danger. This is the second method.

But permanent camps, either for summer or winter, in the neighborhood of an enemy, are fortified with greater care and regularity. After the ground is marked out by the proper officers, each century receives a certain number of feet to entrench. They then range their shields and baggage in a circle about their own colors and, with. out other arms than their swords, open a trench nine, eleven or thirteen feet broad. Or, if they are under great apprehensions of the enemy, they enlarge it to seventeen feet (it being a general rule to observe odd numbers). Within this they construct a rampart with fascines or branches of trees well fastened together with pickets, so that the earth may be better supported. Upon this rampart they raise a parapet with battlements as in the fortifications of a city. The centurions measure the work with rods ten feet long and examine whether every one has properly completed the proportion assigned to him. The tribunes likewise inspect the work and should not leave the place till the whole is finished. And that the workmen may not be suddenly interrupted by the enemy, all the cavalry and that part of the infantry exempted by the privilege of their rank from working, remain in order of battle before the entrenchment to be ready to repel any assault.

The first thing to be done after entrenching the camp, is to plant the ensigns, held by the soldiers in the highest veneration and respect, in their proper places. After this the praetorium is prepared for the general and his lieutenants, and the tents pitched for the tribunes, who have soldiers particularly appointed for that service and to fetch their water, wood, and forage. Then the legions and auxiliaries, cavalry and infantry, have the ground distributed to them to pitch their tents according to the rank of the several corps. Four foot-soldiers of each century and four troopers of each troop are on guard every night. As it seemed impossible for a sentinel to remain a whole night on his post, the watches were divided by the hourglass into four parts, that each man might stand only three hours. All guards are mounted by the sound of trumpet and relieved by the sound of cornet. The tribunes choose proper and trusty men to visit the different posts and report to them whatever they find amiss. This is now a military office and the persons appointed to it are called officers of the rounds.

The cavalry furnish the grand guards at night and the outposts by day. They are relieved every morning and afternoon because of the fatigue of the men and horses. It is particularly incumbent upon the general to provide for the protection of the pastures and of the convoys of grain and other provisions either in camp or garrison, and to secure wood, water and forage against the incursions of the enemy. This can only be effected by posting detachments advantageously in the cines or walled castles on the roads along which the convoys advance. And if no ancient fortifications are to be met with, small forts must be built in proper situations, surrounded with large ditches, for the reception of detachments of horse and foot, so that the convoys will be effectually protected. For an enemy will hardly venture far into a country where he knows his adversary's troops are so disposed as to be ready to encompass him on all sides.

Motives for the Plan of Operations of a Campaign

Readers of this military abridgement will perhaps be impatient for instructions relative to general engagements. But they should consider that a battle is commonly decided in two or three hours, after which no further hopes are left for the worsted army. Every plan, therefore, is to be considered, every expedient tried and every method taken before matters are brought to this last extremity. Good officers decline general engagements where the danger is common, and prefer the employment of stratagem and finesse to destroy the enemy as much as possible in detail and intimidate them without exposing our own forces.

I shall insert some necessary instructions on this head collected from the ancients. It is the duty and interest of the general frequently to assemble the most prudent and experienced officers of the different corps of. the army and consult with them on the state both of his own and the enemy's forces. All overconfidence, as most pernicious in its consequences, must be banished from the deliberations. He must examine which has the superiority in numbers, whether his or the adversary's troops are best armed, which are in the best condition, best disciplined and most resolute in emergencies. The state of the cavalry of both armies must be inquired into, but more especially that of the infantry, for the main strength of an army consists of the latter. With respect to the cavalry, he must endeavor to find out in which are the greatest numbers of archers or of troopers armed with lances, which has the most cuirassiers and which the best horses. Lastly he must consider the field of battle and to judge whether the ground is more advantageous for him or his enemy. If strongest in cavalry, we should prefer plains and open ground; if superior in infantry, we should choose a situation full of enclosures, ditches, morasses and woods, and sometimes mountainous. Plenty or scarcity in either army are considerations of no small importance, for famine, according to the common proverb, is an internal enemy that makes more havoc than the sword. But the most material article is to determine whether it is most proper to temporize or to bring the affair to a speedy decision by action. The enemy sometimes expect an expedition will soon be over; and if it is protracted to any length, his troops are either consumed by want, induced to return home by the desire of seeing their families or, having done nothing considerable in the field, disperse themselves from despair of success. Thus numbers, tired out with fatigue and disgusted with the service, desert, others betray them and many surrender themselves. Fidelity is seldom found in troops disheartened by misfortunes. And in such case an army which was numerous on taking the field insensibly dwindles away to nothing.

It is essential to know the character of the enemy and of their principal officers-whether they be. rash or cautious, enterprising or timid, whether they fight on principle or from chance and whether the nations they have been engaged with were brave or cowardly.

We must know how far to depend upon the fidelity and strength of auxiliaries, how the enemy's troops and our own are affected and which appear most confident of success, a consideration of great effect in raising or depressing the courage of an army. A harangue from the general, especially if he seems under no apprehension himself, may reanimate the soldiers if dejected. Their spirits revive if any considerable advantage is gained either by stratagem or otherwise, if the fortune of the enemy begins to change or if you can contrive to beat some of their weak or poorly-armed detachments.

But you must by no means venture to lead an irresolute or diffident army to a general engagement. The difference is great whether your troops are raw or veterans, whether inured to war by recent service or for some years unemployed. For soldiers unused to fighting for a length of time must be considered in the same light as recruits. As soon as the legions, auxiliaries and cavalry are assembled from their several quarters, it is the duty of a good general to have every corps instructed separately in every part of the drill by tribunes of known capacity chosen for that purpose. He should afterwards form them into one body and train them in all the maneuvers of the line as for a general action. He must frequently drill them himself to try their skill and strength, and to see whether they perform their evolutions with proper regularity and are sufficiently attentive to the sound of the trumpets, the motions of the colors and to his own orders and signals. If deficient in any of these particulars, they must be instructed and exercised till perfect.

But though thoroughly disciplined and complete in their field exercises, in the use of the bow and javelin, and in the evolutions of the line, it is not advisable to lead them rashly or immediately to battle. A favorable opportunity must be watched for, and they must first be prepared by frequent skirmishes and slight encounters. Thus a vigilant and prudent general will carefully weigh in his council the state of his own forces and of those of the enemy, just as a civil magistrate judging between two contending parties. If he finds himself in many respects superior to his adversary, he must by no means defer bringing on an engagement. But if he knows himself inferior, he must avoid general actions and endeavor to succeed by surprises, ambuscades and stratagems. These, when skillfully managed by good generals, have often given them the victory over enemies superior both in numbers and strength.

How to Manage Raw and Undisciplined Troops

All arts and trades whatever are brought to perfection by continual practice. How much more should this maxim, true in inconsiderable matters, be observed in affairs of importance! And how much superior to all others is the art of war, by which our liberties are preserved, our dignities perpetuated and the provinces and the whole Empire itself exist. The Lacedaemonians, and after them the Romans, were so aware of this truth that to this science they sacrificed all others. And the barbarous nations even at this day think only this art worth attention, believing it includes or confers everything else. In short, it is indispensably necessary for those engaged in war not only to instruct them in the means of preserving their own lives, but how to gain the victory over their enemies.

A commander-in-chief therefore, whose power and dignity are so great and to whose fidelity and bravery the fortunes of his countrymen, the defense of their cities, the lives of the soldiers, and the glory of the state, are entrusted, should not only consult the good of the army in general, but extend his care to every private soldier in it. For when any misfortunes happen to those under his command, they are considered as public losses and imputed entirely to his misconduct. If therefore he finds his army composed of raw troops or if they have long been unaccustomed to fighting, he must carefully study the strength, the spirit, the manners of each particular legion, and of each body of auxiliaries, cavalry and infantry. He must know, if possible, the name and capacity of every count, tribune, subaltern and soldier. He must assume the most respectable authority and maintain it by severity. He must punish all military crimes with the greatest rigor of the laws. He must have the character of being inexorable toward offenders and endeavor to give public examples thereof in different places and on different occasions.

Having once firmly established these regulations, he must watch the opportunity when the enemy, dispersed in search of plunder, think themselves in security, and attack them with detachments of tried cavalry or infantry, intermingled with young soldiers, or such as are under the military age. The veterans will acquire fresh experience and the others will be inspired with courage by the advantages such opportunities will give him. He should form ambuscades with the greatest secrecy to surprise the enemy at the passages of rivers, in the rugged passes of mountains, in defiles in woods and when embarrassed by morasses or difficult roads. He should regulate his march so as to fall upon them while taking their refreshments or sleeping, or at a time when they suspect no dangers and are dispersed, unarmed and their horses unsaddled. He should continue these kinds of encounters till his soldiers have imbibed a proper confidence in themselves. For troops that have never been in action or have not for some time been used to such spectacles, are greatly shocked at the sight of the wounded and dying; and the impressions of fear they receive dispose them rather to fly than fight.

If the enemy makes excursions or expeditions, the general should attack him after the fatigue of a long march, fall upon him unexpectedly, or harass his rear. He should detach parties to endeavor to carry off by surprise any quarters established at a distance from the hostile army for the convenience of forage or provisions. F or such measures should be pursued at first as can produce no very bad effects if they should happen to miscarry, but would be of great advantage if attended with success. A prudent general will also try to sow dissention among his adversaries, for no nation, though ever so weak in itself can be completely ruined by its enemies unless its fall be facilitated by its own distraction. In civil dissensions men are so intent on the destruction of their private enemies that they are entirely regardless of the public safety.

One maxim must be remembered throughout this work: that no one should ever despair of effecting what has been already performed. It may be said that our troops for many years past have not even fortified their permanent camps with ditches, ramparts or palisades. The answer is plain. If those precautions had been taken, our armies would never have suffered by surprises of the enemy both by day and night. The Persians, after the example of the old Romans, surround their camps with ditches and, as the ground in their country is generally sandy, they always carry with them empty bags to fill with the sand taken out of the trenches and raise a parapet by piling them one on the other. All the barbarous nations range their carriages round them in a circle, a method which bears some resemblance to a fortified camp. They thus pass their nights secure from surprise.

Are we afraid of not being able to learn from others what they before have learned from us? At present all this is to be found in books only, although formerly constantly practiced. Inquiries are now no longer made about customs that have been so long neglected, because in the midst of peace, war is looked upon as an object too distant to merit consideration. But former instances will convince us that the reestablishment of ancient discipline is by no means impossible, although now so totally lost.

In former ages the art of war, often neglected and forgotten, was as often recovered from books and reestablished by the authority and attention of our generals. Our armies in Spain, when Scipio Africanus took the command, were in bad order and had often been beaten under preceding generals. He soon reformed them by severe discipline and obliged them to undergo the greatest fatigue in the different military works, reproaching them that since they would not wet their hands with the blood of their enemies, they should soil them with the mud of the trenches. In short, with these very troops he afterwards took the city of Numantia and burned it to the ground with such destruction of its inhabitants that not one escaped. In Africa an army, which under the command of Albinus had been forced to pass under the yoke, was by Metellus brought into such order and discipline, by forming it on the ancient model, that they afterwards vanquished those very enemies who had subjected them to that ignominious treatment. The Cimbri defeated the legions of Caepio, Manilus and Silanus in Gaul, but Marius collected their shattered remnants and disciplined them so effectually that he destroyed an innumerable multitude of the Cimbri, Teutones and Ambrones in one general engagement. Nevertheless it is easier to form young soldiers and inspire them with proper notions of honor than to reanimate troops who have been once disheartened.

Preparations for a General Engagement

Having explained the less considerable branches of the art of war, the order of military affairs naturally leads us to the general engagement. This is a conjuncture full of uncertainty and fatal to kingdoms and nations, for in the decision of a pitched battle consists the fulness of victory. This eventuality above all others requires the exertion of all the abilities of a general, as his good conduct on such an occasion gains him greater glory, or his dangers expose him to greater danger and disgrace. This is the moment in which his talents, skill and experience show themselves in their fullest extent.

Formerly to enable the soldiers to charge with greater vigor, it was customary to order them a moderate refreshment of food before an engagement, so that their strength might be the better supported during a long conflict. When the army is to march out of a camp or city in the presence of their enemies drawn up and ready for action, great precaution must be observed lest they should be attacked as they defile from the gates and be cut to pieces in detail. Proper measures must therefore be taken so that the whole army may be clear of the gates and form in order of battle before the enemy's approach. If they are ready before you can have quitted the place, your design of marching out must either be deferred till another opportunity or at least dissembled, so that when they begin to insult you on the supposition that you dare not appear, or think of nothing but plundering or returning and no longer keep their ranks, you may sally out and fall upon them while in confusion and surprise. Troops must never be engaged in a general action immediately after a long march, when the men are fatigued and the horses tired. The strength required for action is spent in the toil of the march. What can a soldier do who charges when out of breath? The ancients carefully avoided this inconvenience, but in later times some of our Roman generals, to say nothing more, have lost their armies by unskillfully neglecting this precaution. Two armies, one tired and spent, the other fresh and in full vigor, are by no means an equal match.

The Sentiments of the Troops Should Be Determined Before Battle

It is necessary to know the sentiments of the soldiers on the day of an engagement. Their confidence or apprehensions are easily discovered by their looks, their words, their actions and their motions. No great dependence is to be placed on the eagerness of young soldiers for action, for fighting has something agreeable in the idea to those who are strangers to it. On the other hand, it would be wrong to hazard an engagement, if the old experienced soldiers testify to a disinclination to fight. A general, however, may encourage and animate his troops by proper exhortations and harangues, especially if by his account of the approaching action he can persuade them into the belief of an easy victory. With this view, he should lay before them the cowardice or unskillfulness of their enemies and remind them of any former advantages they may have gained over them. He should employ every argument capable of exciting rage, hatred and indignation against the adversaries in the minds of his soldiers.

It is natural for men in general to be affected with some sensations of fear at the beginning of an engagement, but there are without doubt some of a more timorous disposition who are disordered by the very sight of the enemy. To diminish these apprehensions before you venture on action, draw up your army frequently in order of battle in some safe situation, so that your men may be accustomed to the sight and appearance of the enemy. When opportunity offers, they should be sent to fall upon them and endeavor to put them to flight or kill some of their men. Thus they will become acquainted with their customs, arms and horses. And the objects with which we are once familiarized are no longer capable of inspiring us with terror.

Choice of the Field of Battle

Good generals are acutely aware that victory depends much on the nature of the field of battle. When you intend therefore to engage, endeavor to draw the chief advantage from your situation. The highest ground is reckoned the best. Weapons thrown from a height strike with greater force; and the party above their antagonists can repulse and bear them down with greater impetuosity, while they who struggle with the ascent have both the ground and the enemy to contend with. There is, however, this difference with regard to place: if you depend on your foot against the enemy's horse, you must choose a rough, unequal and mountainous situation. But if, on the contrary, you expect your cavalry to act with advantage against the enemy's infantry, your ground must indeed be higher, but plain and open, without any obstructions of woods or morasses.

Order of Battle

In drawing up an army in order of battle, three things are to be considered: the sun, the dust and the wind. The sun in your face dazzles the sight: if the wind is against you, it turns aside and blunts the force of your weapons, while it assists those of your adversary; and the dust driving in your front fills the eyes of your men and blinds them. Even the most unskillful endeavor to avoid these inconveniences in the moment of making their dispositions; but a prudent general should extend his views beyond the present; he should take such measures as not to be incommoded in the course of the day by different aspects of the sun or by contrary winds which often rise at a certain hour and might be detrimental during action. Our troops should be so disposed as to have these inconveniences behind them, while they are directly in the enemy's front.

Proper Distances and Intervals

Having explained the general disposition of the lines, we now come to the distances and dimensions. One thousand paces contain a single rank of one thousand six hundred and fifty-six foot soldiers, each man being allowed three feet. Six ranks drawn up on the same extent of ground will require nine thousand nine hundred and ninety-six men. To form only three ranks of the same number will take up two thousand paces, but it is much better to increase the number of ranks than to make your front too extensive. We have before observed the distance between each rank should be six feet, one foot of which is taken up by the men. Thus if you form a body of ten thousand men into six ranks they will occupy thirty-six feet. in depth and a thousand paces in front. By this calculation it is easy to compute the extent of ground required for twenty or thirty thousand men to form upon. Nor can a general be mistaken when thus he knows the proportion of ground for any fixed number of men.

But if the field of battle is not spacious enough or your troops are very numerous, you may form them into nine ranks or even more, for it is more advantageous to engage in close order that to extend your line too much. An army that takes up too much ground in front and too little in depth, is quickly penetrated by the enemy's first onset. After this there is no remedy. As to the post of the different corps in the right or left wing or in the center, it is the general rule to draw them up according to their respective ranks or to distribute them as circumstances or the dispositions of the enemy may require.

Disposition of the Cavalry

The line of infantry being formed, the cavalry are drawn up in the wings. The heavy horse, that is, the cuirassiers and troopers armed with lances, should join the infantry. The light cavalry, consisting of the archers and those who have no cuirasses, should be placed at a greater distance. The best and heaviest horse are to cover the flanks of the foot, and the light horse are posted as abovementioned to surround and disorder the enemy's wings. A general should know what part of his own cavalry is most proper to oppose any particular squadrons or troops of the enemy. For from some causes not to be accounted for some particular corps fight better against others, and those who have defeated superior enemies are often overcome by an inferior force.

If your cavalry is not equal to the enemy's it is proper, after the ancient custom, to intermingle it with light infantry armed with small shields and trained to this kind of service. By observing this method, even though the flower of the enemy's cavalry should attack you, they will never be able to cope with this mixed disposition. This was the only resource of the old generals to supply the defects of their cavalry, and they intermingled the men, used to running and armed for this purpose with light shields, swords and darts, among the horse, placing one of them between two troopers.

Reserves

The method of having bodies of reserves in rear of the army, composed of choice infantry and cavalry, commanded by the supernumerary lieutenant generals, counts and tribunes, is very judicious and of great consequence toward the gaining of a battle. Some should be posted in rear of the wings and some near the center, to be ready to fly immediately to the assistance of any part of the line which is hard pressed, to prevent its being pierced, to supply the vacancies made therein during the action and thereby to keep up the courage of their fellow soldiers and check the impetuosity of the enemy. This was an invention of the Lacedaemonians, in which they were imitated by the Carthaginians. The Romans have since observed it, and indeed no better disposition can be found.

The line is solely designed to repulse, or if possible, break the enemy. If it is necessary to form the wedge or the pincers, it must be done by the supernumerary troops stationed in the rear for that purpose. If the saw is to be formed, it must also be done from the reserves, for if once you begin to draw off men from the line you throw all into confusion. If any flying platoon of the enemy should fall upon your wing or any other part of your army, and you have no supernumerary troops to oppose it or if you pretend to detach either horse or foot from your line for that service by endeavoring to protect one part, you will expose the other to greater danger. In armies not very numerous, it is much better to contract the front, and to have strong reserves. In short, you must have a reserve of good and well-armed infantry near the center to form the wedge and thereby pierce the enemy's line; and also bodies of cavalry armed with lances and cuirasses, with light infantry, near the wings, to surround the flanks of the enemy.

The Post of the General and of the Second and Third in Command

The post of the commander-in-chief is generally on the right between the cavalry and infantry. For from this place he can best direct the motions of the whole army and move elements with the greatest ease wherever he finds it necessary. It is also the most convenient spot to give his orders to both horse and foot and to animate them equally by his presence. It is his duty to surround the enemy's left wing opposed to him with his reserve of horse and light infantry, and attack it in the flank and rear. The second in command is posted in the center of the infantry to encourage and support them. A reserve of good and well-armed infantry is near him and under his orders. With this reserve he either forms the wedge to pierce the enemy's line or, if they form the wedge first, prepares the pincers for its reception. The post of the third in command is on the left. He should be a careful and intrepid officer, this part of the army being difficult to manage and defective, as it were, from its situation in the line. He should therefore have a reserve of good cavalry and active infantry to enable him always to extend his left in such a manner as to prevent its being surrounded.

The war shout should not be begun till both armies have joined, for it is a mark of ignorance or cowardice to give it at a distance. The effect is much greater on the enemy when they find themselves struck at the same instant with the horror of the noise and the points of the weapons.

You must always endeavor to get the start of your enemy in drawing up in order of battle, as you will then have it in your power to make your proper dispositions without obstruction. This will increase the courage of your own troops and intimidate your adversaries. For a superiority of courage seems to be implied on the side of an army that offers battle, whereas troops begin to be fearful who see their enemies ready to attack them. You will also secure another great advantage, that of marching up in order and falling upon them while forming and still in confusion. For part of the victory consists in throwing the enemy into disorder before you engage them.

Maneuvers in Action

An able general never loses a favorable opportunity of surprising the enemy either when tired on the march, divided in the passage of a river, embarrassed in morasses, struggling with the declivities of mountains, when dispersed over the country they think themselves in security or are sleeping in their quarters. In all these cases the adversaries are surprised and destroyed before they have time to put themselves on their guard. But if they are too cautious to give you an opportunity of surprising or ensnaring them, you are then obliged to engage openly and on equal terms. This at present is foreign to the subject. However military skill is no less necessary in general actions than in carrying on war by subtlety and stratagem.

Your first care is to secure your left wing from being surrounded by the enemy's numbers or attacked in flank or rear by flying platoons, a misfortune that often happens. Nor is your right to be neglected, though less frequently in danger. There is only one remedy for this: to wheel back your wing and throw it into a circular position. By this evolution your soldiers meet the enemy on the quarter attacked and defend the rear of their comrades. But your best men should be posted on the angles of the flanks, since it is against them the enemy make their principal efforts.

There is also a method of resisting the wedge when formed by the enemy. The wedge is a disposition of a body of infantry widening gradually toward the base and terminating in a point toward the front. It pierces the enemy's line by a multitude of darts directed to one particular place. The soldiers call it the swine's head. To oppose this disposition, they make use af another called the pincers, resembling the letter V, composed of a body of men in close order. It receives the wedge, inclosing it on both sides, and thereby prevents it from penetrating the line.

The saw is another disposition formed of resolute soldiers drawn up in a straight line advanced into the front against the enemy, to repair any disorder. The platoon is a body of men separated from the line, to hover on every side and attack the enemy wherever they find opportunity. And against this is to be detached a stronger and more numerous platoon.

Above all, a general must never attempt to alter his dispositions or break his order of battle during the time of action, for such an alteration would immediately Occasion disorder and confusion which the enemy would not fail to improve to their advantage.

Various Formations for Battle

An army may be drawn up for a general engagement in seven different formations. The first formation is an oblong square of a large front, of common use both in ancient and modern times, although not thought the best by various judges of the service, because an even and level plain of an extent sufficient to contain its front cannot always be found, and if there should be any irregularity or hollow in the line, it is often pierced in that part. Besides, an enemy superior in number may surround either your right or left wing, the consequence of which will be dangerous, unless you have a reserve ready to advance and sustain his attack. A general should make use of this disposition only when his forces are better and more numerous than the enemy's, it being thereby in his power to attack both the flanks and surround them on every side.

The second and best disposition is the oblique. For although your army consists of few troops, yet good and advantageously posted, it will greatly contribute to your obtaining the victory, notwithstanding the numbers and bravery of the enemy. It is as follows: as the armies are marching up to the attack, your left wing must be kept back at such a distance from the enemy's right as to be out of reach of their darts and arrows. Your right wing must advance obliquely upon the enemy's left, and begin the engagement. And you must endeavor with your best cavalry and infantry to surround the wing with which you are engaged, make it give way and fall upon the enemy in the rear. If they once give ground and the attack is properly seconded, you will undoubtedly gain the victory, while your left wing, which continued at a distance, will remain untouched. An army drawn up in this manner bears some resemblance to the letter A or a mason's level. If the enemy should be beforehand with you in this evolution, recourse must be had to the supernumerary horse and foot posted as a reserve in the rear, as I mentioned before. They must be ordered to support your left wing. This will enable you to make a vigorous resistance against the artifice of the enemy.

The third formation is like the second, but not so good, as it obliges you to begin the attack with your left wing on the enemy's right. The efforts of soldiers on the left are weak and imperfect from their exposed and defective situation in the line. I will explain this formation more clearly. Although your left wing should be much better than your right, yet it must be reinforced with some of the best horse and foot and ordered to commence the acnon with the enemy's right in order to disorder and surround it as expeditiously as possible. And the other part of your army, composed of the worst troops, should remain at such a distance from the enemy's left as not to be annoyed by their darts or in danger of being attacked sword in hand. In this oblique formation care must be taken to prevent the line being penetrated by the wedges of the enemy, and it is to be employed only when the enemy's right wing is weak and your greatest strength is on your left.

The fourth formation is this: as your army is marching to the attack in order of battle and you come within four or five hundred paces of the enemy, both your wings must be ordered unexpectedly to quicken their pace and advance with celerity upon them. When they find themselves attacked on both wings at the same time, the sudden surprise may so disconcert them as to give you an easy victory. But although this method, if your troops are very resolute and expert, may ruin the enemy at once, yet it is hazardous. The general who attempts it is obliged to abandon and expose his center and to divide his army into three parts. If the enemy are not routed at the first charge, they have a fair opportunity of attacking the wings which are separated from each other and the center which is destitute of assistance.

The fifth formation resembles the fourth but with this addition: the light infantry and the archers are formed before the center to cover it from the attempts of the enemy. With this precaution the general may safely follow the above mentioned method and attack the enemy's left wing with his right, and their right with his left. If he puts them to flight, he gains an immediate victory, and if he fails of success his center is in no danger, being protected by the light infantry and archers.

The sixth formation is very good and almost like the second. It is used when the general cannot depend either on the number or courage of his troops. If made with judgment, notwithstanding his inferiority, he has often a good chance for victory. As your line approaches the enemy, advance your right wing against their left and begin the attack with your best cavalry and infantry. At the same time keep the rest of the army at a great distance from the enemy's right, extended in a direct line like a javelin. Thus if you can surround their left and attack it in flank and rear, you must inevitably defeat them. It is impossible for the enemy to draw off reinforcements from their right or from their center to sustain their left in this emergency, since the remaining part of your army is extended and at a great distance from them in the form of the letter L. It is a formation often used in an action on a march.

The seventh formation owes its advantages to the nature of the ground and will enable you to oppose an enemy with an army inferior both in numbers and goodness, provided one of your flanks can be covered either with an eminence, the sea, a river, a lake, a city, a morass or broken ground inaccessible to the enemy. The rest of the army must be formed, as usual, in a straight line and the unsecured flank must be protected by your light troops and all your cavalry. Sufficiently defended on one side by the nature of the ground and on the other by a double support of cavalry, you may then safely venture on action.

One excellent and general rule must be observed. If you intend to engage with your right wing only, it must be composed of your best troops. And the same method must be taken with respect to the left. Or if you intend to penetrate the enemy's line, the wedges which you form for that purpose before your center, must consist of the best disciplined soldiers. Victory in general is gained by a small number of men. Therefore the wisdom of a general appears in nothing more than in such choice of disposition of his men as is most consonant with reason and service.

The Flight of an Enemy Should Not Be Prevented, But Facilitated

Generals unskilled in war think a victory incomplete unless the enemy are so straightened in their ground or so entirely surrounded by numbers as to have no possibility of escape. But in such situation, where no hopes remain, fear itself will arm an enemy and despair inspires courage. When men find they must inevitably perish, they willingly resolve to die with their comrades and with their arms in their hands. The maxim of Scipio, that a golden bridge should be made for a flying enemy, has much been commended. For when they have free room to escape they think of nothing but how to save themselves by flight, and the confusion becoming general, great numbers are cut to pieces. The pursuers can be in no danger when the vanquished have thrown away their arms for greater haste. In this case the greater the number of the flying army, the greater the slaughter. Numbers are of no signification where troops once thrown into consternation are equally terrified at the sight of the enemy as at their weapons. But on the contrary, men when shut up, although weak and few in number, become a match for the enemy from this very reflection, that they have no resource but in despair.

"The conquer'd's safety is, to hope for none."

Manner of Conducting a Retreat

Having gone through the various particulars relative to general actions, it remains at present to explain the manner of retreating in presence of the enemy. This is an operation, which, in the judgment of men of greatest skill and experience, is attended with the utmost hazard. A general certainly discourages his own troops and animates his enemies by retiring out of the field without fighting. Yet as this must sometimes necessarily happen, it will be proper to consider how to perform it with safety.

In the first place your men must not imagine that you retire to decline an action, but believe your retreat an artifice to draw the enemy into an ambuscade or more advantageous position where you may easier defeat them in case they follow you. For troops who perceive their general despairs of success are prone to flight. You must be cautious lest the enemy should discover your retreat and immediately fall upon you. To avoid this danger the cavalry are generally posted in the front of the infantry to conceal their motions and retreat from the enemy. The first divisions are drawn off first, the others following in their turns. The last maintain their ground till the rest have marched off, and then file off themselves and join them in a leisurely and regular succession. Some generals have judged it best to make their retreat in the night after reconnoitering their routes, and thus gain so much ground that the enemy, not discovering their departure till daybreak, were not able to come up with them. The light infantry was also sent forward to possess the eminences under which the army might instantly retire with safety; and the enemy, in case they pursued, be exposed to the light infantry, masters of the heights, seconded by the cavalry.

A rash and inconsiderate pursuit exposes an army to the greatest danger possible, that of falling into ambuscades and the hands of troops ready for their reception. For as the temerity of an army is increased and their caution lessened by the pursuit of a flying enemy, this is the most favorable opportunity for such snares. The greater the security, the greater the danger. Troops, when unprepared, at their meals, fatigued after a march, when their horses are feeding, and in short, when they believe themselves most secure, are generally most liable to a surprise. All risks of this sort are to be carefully avoided and all opportunities taken of distressing the enemy by such methods. Neither numbers nor courage avail in misfortunes of this nature.

A general who has been defeated in a pitched battle, although skill and conduct have the greatest share in the decision, may in his defense throw the blame on fortune. But if he has suffered himself to be surprised or drawn into the snares of his enemy, he has no excuse for his fault, because he might have avoided such a misfortune by taking proper precautions and employing spies on whose intelligence he could depend.

When the enemy pursue a retreating foe, the following snare is usually laid. A small body of cavalry is ordered to pursue them on the direct road. At the same time a strong detachment is secretly sent another way to conceal itself on their route. When the cavalry have overtaken the enemy, they make some feint attacks and retire. The enemy, imagining the danger past, and that they have escaped the snare, neglect their order and march without regularity. Then the detachment sent to intercept them, seizing the opportunity, falls upon them unexpectedly and destroys them with ease.

Many generals when obliged to retreat through woods send forward parties to seize the defiles and difficult passes, to avoid ambuscades and block the roads with barricades of felled trees to secure themselves from being pursued and attacked in the rear. In short both sides have equal opportunities of surprising or laying ambuscades on the march. The army which retreats leaves troops behind for that purpose posted in convenient valleys or mountains covered with woods, and if the enemy falls into the snare, it returns immediately to their assistance. The army that pursues detaches different parties of light troops to march ahead through by-roads and intercepts the enemy, who are thus surrounded and attacked at once in front and rear. The flying army may return and fall on the enemy while asleep in the night. And the pursuing army may, even though the distance is great, surprise the adversary by forced marches. The former endeavor may be at the crossing of a river in order to destroy such part of the enemy's army as has already crossed. The pursuers hasten their march to fall upon those bodies of the enemy that have not yet crossed.

Armed Chariots and Elephants

The armed chariots used in war by Antiochus and Mithridates at first terrified the Romans, but they afterwards made a jest of them. As a chariot of this sort does not always meet with plain and level ground, the least obstruction stops it. And if one of the horses be either killed or wounded, it falls into the enemy's hands. The Roman soldiers rendered them useless chiefly by the following contrivance: at the instant the engagement began, they strewed the field of battle with caltrops, and the horses that drew the chariots, running full speed on them, were infallibly destroyed.

A caltrop is a machine composed of four spikes or points arranged so that in whatever manner it is thrown on the ground, it rests on three and presents the fourth upright.

Elephants by their vast size, horrible noise and the novelty of their form are at first very terrible both to men and horses. Pyrrhus first used them against the Romans in Lucania. And afterwards Hannibal brought them into the field in Africa. Antiochus in the east and Jugurtha in Numidia had great numbers. Many expedients have been used against them. In Lucania a centurion cut off the trunk of one with his sword. Two soldiers armed from head to foot in a chariot drawn by two horses, also covered with armor, attacked these beasts with lances of great length. They were secured by their armor from the archers on the elephants and avoided the fury of the animals by the swiftness of their horses. Foot soldiers completely armored, with the addition of long iron spikes fixed on their arms, shoulders and helmets, to prevent the elephant from seizing them with his trunk, were also employed against them.

But among the ancients, the velites usually engaged them. They were young soldiers, lightly armed, active and very expert in throwing their missile weapons on horseback. These troops kept hovering round the elephants continually and killed them with large lances and javelins. Afterwards, the soldiers, as their apprehensions decreased, attacked them in a body and, throwing their javelins together, destroyed them by the multitude of wounds. Slingers with round stones from the fustibalus and sling killed both the men who guided the elephants and the soldiers who fought in the towers on their backs. This was found by experience to be the best and safest expedient. At other times on the approach of these beasts, the soldiers opened their ranks and let them pass through. When they got into the midst of the troops, who surrounded them on all sides, they were captured with their guards unhurt.

Large balistae, drawn on carriages by two horses or mules, should be placed in the rear of the line, so that when the elephants come within reach they may be transfixed with the darts. The balistae should be larger and the heads of the darts stronger and broader than usual, so that the darts may be thrown farther, with greater force and the wounds be proportioned to the bodies of the beasts. It was proper to describe these several methods and contrivances employed against elephants, so that it may be known on occasion in what manner to oppose those prodigious animals.

Resources in Case of Defeat

If while one part of your army is victorious the other should be defeated, you are by no means to despair, since even in this extremity the constancy and resolution of a general may recover a complete victory. There are innumerable instances where the party that gave least way to despair was esteemed the conqueror. For where losses and advantages seem nearly equal, he is reputed to have the superiority who bears up against his misfortunes with greatest resolution. He is therefore to be first, if possible, to seize the spoils of the slain and to make rejoicings for the victory. Such marks of confidence dispirit the enemy and redouble your own courage.

Yet notwithstanding an entire defeat, all possible remedies must be attempted, since many generals have been fortunate enough to repair such a loss. A prudent officer will never risk a general action without taking such precautions as will secure him from any considerable loss in case of a defeat, for the uncertainty of war and the nature of things may render such a misfortune unavoidable. The neighborhood of a mountain, a fortified post in the rear or a resolute stand made by a good body of troops to cover the retreat, may be the means of saving the army.

An army after a defeat has sometimes rallied, returned on the enemy, dispersed him by pursuing in order and destroyed him without difficulty. Nor can men be in a more dangerous situation than, when in the midst of joy after victory, their exultation is suddenly converted into terror. Whatever be the event, the remains of the army must be immediately assembled, reanimated by suitable exhortations and furnished with fresh supplies of arms. New levies should immediately be made and new reinforcements provided. And it is of much the greatest consequence that proper opportunities should be taken to surprise the victorious enemies, to draw them into snares and ambuscades and by this means to recover the drooping spirits of your men. Nor will it be difficult to meet with such opportunities, as the nature of the human mind is apt to be too much elated and to act with too little caution in prosperity. If anyone should imagine no resource is left after the loss of a battle, let him reflect on what has happened in similar cases and he will find that they who were victorious in the end were often unsuccessful in the beginning.

General Maxims

- It is the nature of war that what is beneficial to you is detrimental to the enemy and what is of service to him always hurts you. It is therefore a maxim never to do, or to omit doing, anything as a consequence of his actions, but to consult invariably your own interest only. And you depart from this interest whenever you imitate such measures as he pursues for his benefit. For the same reason it would be wrong for him to follow such steps as you take for your advantage.
- The more your troops have been accustomed to camp duties on frontier stations and the more carefully they have been disciplined, the less danger they will be exposed to in the field.
- Men must be sufficiently tried before they are led against the enemy.
- It is much better to overcome the enemy by famine, surprise or terror than by general actions, for in the latter instance fortune has often a greater share than valor. Those designs are best which the enemy are entirely ignorant of till the moment of execution. Opportunity in war is often more to be depended on than courage.
- To debauch the enemy's soldiers and encourage them when sincere in surrendering themselves, is of especial service, for an adversary is more hurt by desertion than by slaughter.

- It is better to have several bodies of reserves than to extend your front too much.
- A general is not easily overcome who can form a true judgment of his own and the enemy's forces.
- Valor is superior to numbers.
- The nature of the ground is often of nmore consequence than courage.
- Few men are born brave; many become so through care and force of discipline.
- An army is strengthened by labor and enervated by idleness.
- Troops are not to be led to battle unless confident of success.
- Novelty and surprise throw an enemy into consternation; but common incidents have no effect.
- He who rashly pursues a flying enemy with troops in disorder, seems inclined to resign that victory which he had before obtained.
- An army unsupplied with grain and other necessary provisions will be vanquished without striking a blow.
- A general whose troops are superior both in number and bravery should engage in the oblong square, which is the first formation.
- He who judges himself inferior should advance his right wing obliquely against the enemy's left. This is the second formation.
- If your left wing is strongest, you must attack

the enemy's right according to the third formation.

- The general who can depend on the discipline of his men should begin the engagement by attacking both the enemy's wings at once, the fourth formation.
- He whose light infantry is good should cover his center by forming them in its front and charge both the enemy's wings at once. This is the fifth formation.
- He who cannot depend either on the number or courage of his troops, if obliged to engage, should begin the action with his right and endeavor to break the enemy's left, the rest of his army remaining formed in a line perpendicular to the front and extended to the rear like a javelin. This is the sixth formation.
- If your forces are few and weak in comparison to the enemy, you must make use of the seventh formation and cover one of your flanks either with an eminence, a city, the sea, a river or some protection of that kind.
- A general who trusts to his cavalry should choose the proper ground for them and employ them principally in the action.
- He who depends on his infantry should choose a situation most proper for them and make most use of their service.
- When an enemy's spy lurks in the camp, order all your soldiers in the day time to their tents, and he will instantly be apprehended.

- On finding the enemy has notice of your designs, you must immediately alter your plan of operations.
- Consult with many on proper measures to be taken, but communicate the plans you intend to put in execution to few, and those only of the most assured fidelity; or rather trust no one but yourself.
- Punishment, and fear thereof, are necessary to keep soldiers in order in quarters; but in the field they are more influenced by hope and rewards.
- Good officers never engage in general actions unless induced by opportunity or obliged by necessity.
- To distress the enemy more by famine than the sword is a mark of consummate skill.
- Many instructions might be given with regard to the cavalry. But as this branch of the service has been brought to perfection since the ancient writers and considerable improvements have been made in their drills and maneuvers, their arms, and the quality and management of their horses, nothing can be collected from their works. Our present mode of discipline is sufficient.
- Dispositions for action must be carefully concealed from the enemy, lest they should counteract them and defeat your plans by proper expedients.
- This abridgment of the most eminent military writers, invincible Emperor, contains the

maxims and instructions they have left us, approved by different ages and confirmed by repeated experience. The Persians admire your skill in archery; the Huns and Alans endeavor in vain to imitate your dexterity in horsemanship; the Saracens and Indians cannot equal your activity in the hunt; and even the masters at arms pique themselves on only part of that knowledge and expertness of which you give so many instances in their own profession. How glorious it is therefore for Your Majesty with all these qualifications to unite the science of war and the art of conquest, and to convince the world that by Your conduct and courage You are equally capable of performing the duties of the soldier and the general!

• An instrument with four points so designed that when any three of them are on the ground the fourth projects upward. These are extensively used today for antitank barriers.

e|LIVE

listen|imagine|view|experience

AUDIO BOOK DOWNLOAD INCLUDED WITH THIS BOOK!

In your hands you hold a complete digital entertainment package. Besides purchasing the paper version of this book, this book includes a free download of the audio version of this book. Simply use the code listed below when visiting our website. Once downloaded to your computer, you can listen to the book through your computer's speakers, burn it to an audio CD or save the file to your portable music device (such as Apple's popular iPod) and listen on the go!

How to get your free audio book digital download:

1. Visit www.tatepublishing.com and click on the e|LIVE logo on the home page.
2. Enter the following coupon code: dbd6-2dfc-e838-3db9-fe34-2b1f-df0e-efdd
3. Download the audio book from your e|LIVE digital locker and begin enjoying your new digital entertainment package today!